THE
SENSITIVES

The Rise of
Environmental Illness
—— and the ——
Search for America's
Last Pure Place

OLIVER BROUDY

Simon & Schuster
New York   London   Toronto   Sydney   New Delhi

Simon & Schuster
1230 Avenue of the Americas
New York, NY 10020

Some names and identifying characteristics have been changed.

First Simon & Schuster hardcover edition July 2020

SIMON & SCHUSTER and colophon are registered
trademarks of Simon & Schuster, Inc.

For information about special discounts for bulk purchases,
please contact Simon & Schuster Special Sales at 1-866-506-1949
or business@simonandschuster.com.

The Simon & Schuster Speakers Bureau can bring authors to your
live event. For more information or to book an event, contact the
Simon & Schuster Speakers Bureau at 1-866-248-3049 or visit our
website at www.simonspeakers.com.

Interior design by Lewelin Polanco

Manufactured in the United States of America

10 9 8 7 6 5 4 3 2 1

Library of Congress Cataloging-in-Publication Data
  Names: Broudy, Oliver, author.
  Title: The sensitives / Oliver Broudy.
  Description: New York: Simon & Schuster, [2020] | Includes bibliographical refer-
ences and index. | Summary: A compelling exploration of the mysteries of environ-
mental toxicity and the community of "sensitives"—people with powerful, puzzling
symptoms resulting from exposure to chemicals, fragrances, and cell phone signals,
that have no effect on "normals."—Provided by publisher.
  Identifiers: LCCN 2019050181 | ISBN 9781982128500 (hardcover) |
ISBN 9781982128524 (paperback) | ISBN 9781982128548 (ebook)
  Subjects: LCSH: Environmentally induced diseases—United States—Case
studies. | Environmental health—United States—Case studies. | Environmental
toxicology—United States—Case studies.
  Classification: LCC RB152.5 .B756 2020 | DDC 616.9/800922—dc23
  LC record available at https://lccn.loc.gov/2019050181

ISBN 978-1-9821-2850-0
ISBN 978-1-9821-2854-8 (ebook)

for B.F.

Contents

DAY 3

DAY 4

DAY 5

DAY 6

THE
SENSITIVES

CHAPTER 1

The Disappearance of Brian Welsh

B rian Welsh was not acclaimed in any of the usual ways. He was no football star or academic prodigy. He was no senator's son. In fact his dad was a verbally abusive factory worker who, thankfully perhaps, died young from emphysema.

Growing up in Michigan, Brian was well liked by his peers, one of those good-natured extroverts for whom clowning and raillery came easy. After high school, he studied journalism at the local community college, drove a forklift at a lumberyard to pay the bills, and eventually got married. When the lumberyard job disappeared he trained to become a respiratory therapist. For a boy from Muskegon, life wasn't looking half-bad.

And then, within the space of a few years, Brian found himself cast out of his own life—excommunicated by friends and family, unable to work, divorced by his wife.

It was the kind of theatrical downfall that, in a different era, would sometimes follow the breaking of some dire taboo, like getting pregnant out of wedlock, or advocating communism at the neighborhood potluck.

Only in this case it was because Brian had gotten sick, and the kind of sick he got made sense to no one. Because of this he was thrust past sick to some farther margin of society's firelight, an outer dark where order and reason gave way to uncertainty and confusion.

When it began he had no idea what was wrong with him, only that he felt tired, and could no longer tolerate certain foods. But within months he found himself too fatigued to go out at night or play sports. Then he began reacting to specific triggers in his environment. Perfume set his heart racing. Paint fumes fogged his brain. Meanwhile the list of foods he could tolerate grew ever shorter.

He tried everything he could think of—acupuncture, chiropractors, supplements, methylation, even a fecal transplant courtesy of his wife—a final favor before the divorce. Nothing worked. Eventually he reached a point where he could no longer even stand to be indoors.

The last thread parted in the fall of 2014 when the Polar Vortex swept down from Canada, rousting Brian from where he was sheltering on his aunt's back porch. With that he was adrift, without anchor or direction.

The swiftness, the methodical *thoroughness*, of Brian's unmaking carried a certain Jobean trauma. A merciless humbling that stripped away everything, from material assets like clothes and housing to conceptual assets like threat awareness and body knowledge—as well as all the relationships that gave his life meaning.

His one solace was years in coming. But eventually he figured out that he was not the only one. There were others. Thousands. Millions.

They called themselves "sensitives," because that is what they were. Environmental factors that to others were undetectable or benign—hair spray, for instance—were to them noxious or intolerable. Almost any chemical or toxin could be a trigger, but also cell phone radiation and some types of mold. Even the slightest exposure to the most ordinary products could be debilitating. A whiff of flavored coffee could shut them down for a week. So could a scented dog poo bag, a Wi-Fi router, ice cream with preservatives. In the beginning, the list of triggers tended to be very limited and specific, but as time went on the list could grow much longer.

The most common symptoms were fatigue, brain fog, muscle aches, and memory difficulties, but could include almost anything. And efforts to avoid them could lead to strikingly aberrant behavior. Hanging mail from a clothesline for a week to "off-gas" before daring to read it, for instance. Shaving facial hair to reduce susceptibility to airborne pesticides. Yanking out teeth. One sensitive, leery of chemically saturated waiting rooms, convinced her gynecologist to conduct his examination in the backseat of her car. Another, leery of laundry detergent, went nine years without clean sheets.

The phenomenon went by many names, including Multiple Chemical Sensitivity, Idiopathic Environmental Intolerance, and Total Allergy Syndrome, but among them all I had found Environmental Illness, or EI, to be the most impartial and inclusive. First recognized in 1962, EI was believed to afflict as many as 42 million people nationwide. Over the last decade alone the prevalence had increased over 300 percent, with as much as 30 percent of the population experiencing some level of hypersensitivity.

No one was born with EI. Usually it commenced with a single massive toxic exposure, like a chemical spill, or else years of low-level exposures, as with a job at a plastics factory. Then came a bewildering period in which the victim struggled to make sense of a world where peril lurked around every corner. In the more extreme cases, this often involved exhausting their life savings on desperate treatments ranging from nutritional supplements to exorcism. Eventually, however, the attempt to cure the EI would be abandoned and the focus would shift to finding someplace in the world they could catch their breath and try to remember who they were. Some ended up in houses wallpapered in tinfoil, driving thirty-year-old Benzes with gutted electrical systems and solar panels on the roofs. Others didn't leave their homes for years, or would only venture out wearing an industrial-grade respirator.

The endless search for refuge took its own toll. One sensitive I talked to had bought and sold four houses, losing $20,000 to $30,000 each time— and never got to live in any of them. Another said his EI had cost him $850,000. Those who had been financially shattered by the disease could be found living out of tents in national parks, or cars in Walmart parking lots, fleeing from one oasis to the next as the seasons changed—or even the winds, which often carried pollutants from nearby factories or farms.

"Sensitives," they called themselves—but also "the new refugees," or "runners"—this last a nod to the imperative that consumed the lives of so many as they fled one toxic exposure after another. One guy I'd spoken to claimed to have slept in 240 places in eight years. Another told me the only place he felt like his pre-EI self was halfway up Mount McKinley.

"A glacier on the side of a mountain is for me like paradise," he said. "It's about the only place in the world where my symptoms are gone completely."

Talking with these folks you got the sense of a vast struggle invisibly underway all across the nation. A steady, Atlas-like exertion just to maintain hope.

"It's almost like being a pregnant woman," a sensitive named David Reeves told me. "A pregnant woman has all these crazy symptoms and she's always kinda sick, and that's what it's like. Except, after nine months, man, it just keeps going."

Worst of all was that no one could agree on what EI was—or whether it even existed. The mechanism of action had been theorized to include everything from immunological dysregulation to schizophrenia, but no one really had any idea. It was often associated with anxiety and depression—but whether it was a result of these disorders or a cause was unclear; nor whether the association was any greater than could be found in other chronic diseases. It appeared to be limited to Western industrialized countries—or the reporting of it was, anyway—but the epidemiology suggested no obvious pattern: radiologists in New Zealand and sheep dippers in Great Britain; log cabin dwellers in Germany and hospital workers in Nova Scotia. It was even found among employees of the Environmental Protection Agency itself, after their headquarters was renovated and new carpeting installed.

It seemed like it could happen to anyone. In fact, David, the sensitive who cited the pregnancy comparison, could very easily have been me. David was a New York publishing guy whose life fell apart after his apartment got sprayed for bedbugs. The next thing he knew he was living out of a $30 tent at the southern end of the Rincon Mountains, passing his days reading Trollope on a cracked iPad and listening to Gila monsters hump in the driveway.

"You have no possessions," he told me, "you have the clothes on your back and a five-gallon plastic jug for carrying water. Like, that's it. And an iPad. And that's your life."

The syndrome's only other distinguishing points were that it was more prevalent among adults, less among youth and seniors; and more prevalent among women. But then the same could be said of a host of other elusive ailments, like myalgic encephalomyelitis, fibromyalgia, rheumatoid arthritis, and chronic Lyme disease.

Consequently, neither the American Medical Association nor the Centers for Disease Control and Prevention accorded EI clinical validity. As far as the entire medical establishment was concerned, it was at best quasi-real.

For sensitives, this itself was a constant source of anguish, because at the end of the day their suffering was not considered legit. Laughed at by strangers, dismissed by medical professionals, renounced as "crazy" by their own friends and family, they were rendered unpersons, denied the right to tell their own story.

One sensitive referred to EI as "the divorce disease." Another, a former psychotherapist, told me that he often wished he'd gotten cancer instead.

"At least then people believe you," he said.

Given the institutional conservatism of organizations like the AMA and the CDC, I was inclined to give sensitives the benefit of the doubt. The threat of chemicals and toxins was hardly unfounded, after all. Our world was awash with them. Even the AMA quietly admitted that EI could not be dismissed as merely "psychogenic." The CDC had made no such admission, but they did make a point of forbidding air fresheners, perfumes, scented hand lotion, even urinal pucks in the workplace due to their "detrimental effects on the health of chemically sensitive co-workers."

But even if the AMA and CDC weren't deserving of a little skepticism I had no wish to contribute to the pain of sensitives by dismissing their experience out of hand. On the contrary, the deep, existential tenor of their suffering filled me with something like awe. It was an awe they often felt themselves.

"At times you find yourself left with only the moment you're in," Brian told me.

At that point suicide came to seem like a relatively trivial matter, like stubbing out the remains of a bonfire that had already dwindled to a spark. Brian had told me exactly how he would do it, gunning his pickup off a certain stretch of road just north of Phoenix. The road there was conveniently devoid of guardrails and the drop was over three hundred feet. With all his earthly belongings packed in the truck bed behind him, he said, there'd be nothing for anyone to clean up. The thoughtfulness of this detail was typical of Brian, and very much in keeping with the code of the National Parks where he had often sought refuge: take nothing but memories, leave nothing but footprints.

Maybe it was this closeness to extinction that endowed Brian with his uncanny transparency. In losing his place in society he had also lost all the affectation and obfuscation that went with it. With this stuff out of the way, talking to Brian could sometimes be disorienting as he favored you with what felt like unearned trust. If you could accept it at face value the effect was oddly elating—like a gift. But it could also be a little disturbing, as if you'd just entered a room in which the usual rules of physics no longer applied.

In one of our few phone conversations Brian had told me a story about being menaced by some local hunters who didn't like the idea of him camping in the middle of their woods. They rolled up in their pickups cursing and threatening, and later took to stalking him at night. Yet Brian was weirdly unafraid. In fact, whenever the hunters appeared he would walk right up and engage them in polite conversation, like the forest's own maître d'. This exasperated the hunters even more. But for Brian the animosity never registered. In embracing his own vulnerability, fear had somehow ceased to be relevant. This, too, had been taken from him— like everything else. Everything but the moment. He had become like a creature of cellophane. Not empty but translucent.

It was this same translucency that had made Brian a prominent figure in the underground EI communities flourishing online. From the extremity of his circumstance, dipping in and out of lucidity and sharing stories of his own travails—crisscrossing the country, regularly being mistaken for a vagrant, a pederast, a meth dealer, a thief—he captured and validated what everyone was feeling. That frazzled, half-mad desperation. That shattered spirit pushed way past its limits but still hanging on.

In this, Brian had become an emotional conduit for the community as a whole, modeling his translucency for a people who had become tragically obscure to themselves. His stories bespoke a collective trauma—even as he struggled every day to find and voice new reasons to stay alive. Taken together, his countless Facebook posts amounted to a running commentary on the nature of suffering itself.

"It's not about finally finding out who you are," he would say, "it's about finding out what it feels like to truly be in need. To need desperately. That's what truly changes a person."

It was for this that he was so often called "courageous," "an inspiration," "a treasure," even "a mystic." It was for this that he was told, over and over again, to "write a book." But then every so often when I talked with Brian he would say something so . . . difficult to accommodate that skepticism became impossible to suppress. Like when he spoke of moving his sleeping bag every night to avoid the ground he had polluted simply by lying on it. It was no longer a question of avoiding the contaminant, in other words. In his mind, he had *become* the contaminant. Or the contaminant had become him. It sounded like a story of possession. Thus the exorcism, I suppose.

It happened over and over again, this peculiar, baffled moment—and not just with Brian. Something about it was endemic to the condition. One sensitive described passing out cold when a woman walked past him at the airport trailing a cloud of perfume. Another claimed, in all seriousness, that newsprint made her ass itch. Most memorable of all was the woman convinced that a spasm in her toe was caused by a passing blimp.

The anecdotes carried a whiff of madness. Yet the distress the sensitives expressed was clearly real. In these moments they seemed to me both mad and not mad at the same time, like a particle that is also a wave—depending on how you looked at it. The effect on the mind was wrenching, like those shifts in perspective when the young woman suddenly turned into a crone, or the rabbit into a duck. Except worse because everything that went along with each perspective—how you felt about chemicals and pollution, how you felt about mainstream medicine, about government regulation, about suffering itself, not to mention the actual human being you were talking to—got wrenched with it. How you looked at them also changed how you looked at you.

Such was the riddle that had begun to preoccupy me when I learned one day that Brian had gone missing. The last anyone had heard he was somewhere on the North Rim of the Grand Canyon, $40,000 in debt, and maxing out the last of his credit cards to buy supplements from Amazon. Now, after days of silence and mounting concern from his online community, it seemed that he, like so many others of his kind, had finally succumbed to that outer dark where he had been thrust.

The wisest choice was to ignore this development—avert the eyes and carry on. But I had done that. Somehow it felt like I was always doing that.

I would not avert my eyes from Brian. I would find him, if he could possibly be found.

DAY
1

CHAPTER 2

On the Run

James was my Virgil. He knew the roads, having spent years circling through them. In the last two years alone he had traveled 44,000 miles, only once leaving the country, and then just for a day, hopping over the border to Calgary before being repulsed by a bad hotel and ambient industrial fumes.

James was currently in Aspen, about to embark on a road trip to L.A., where he had an appointment with a doctor in whom he'd placed his last hopes for a cure. The road to L.A. passed very close to Brian's last known location, and I asked James if I could hitch a ride, figuring the days on the road would give me time and mental space to sort through EI's many puzzling contradictions. At the time I could not have anticipated that the most puzzling contradiction of them all would turn out to be James himself.

The first hint of James's peculiar character came with his energetic acceptance of my proposal. For what kind of person invites a complete stranger into his car for a multiday road trip, no questions asked? After all, not everyone with his condition was as open and trusting as Brian. In my limited experience, most sensitives were either cagey to the point of paranoia or else much like the rest of us, torn between a desire to be understood and an opposing desire to protect themselves and avoid further pain.

Perhaps James felt that in talking to me he could help others. Certainly I had encountered this motivation before in previous investigations. Or maybe he was just naive. You saw this now and then among the super-wealthy, a friendly myopia that started out feeling inclusive until you realized that it was premised on the failure to understand that not everyone owns their own plane.

With his green polo shirt and Prada sunglasses the overall impression was jock prep, high school alpha. Like a Ralph Lauren model. He had sounded saner than most on the phone—aside from his oddly hesitant manner of speaking, just shy of a stutter, which often left the conversation suspended mid-sentence as he groped for a word. Possibly this signaled some sort of cognitive disability due to EI. Then again, maybe the traffic was so bad (he'd been driving at the time) that he just couldn't concentrate.

The most likely explanation for James's willingness to help was simple boredom. Like a trucker, he was always on the road, and the road could be monotonous. This was one of the things I liked about James, actually: how literally unsettled he was. It seemed right, somehow, given the state of the country, with a TV buffoon leading in the presidential race, and newspaper headlines which every day offered new reasons to pack up and flee. I never had much patience for the folks who were always threatening to move to Canada. Somehow they never quite got around to it. But even for a stoic like myself, holding still as the world worsened around me had begun to feel unusually stressful. Endless mobility seemed a decent compromise. A way of making a semicitizen of oneself, detaching from a land that was becoming less habitable every day.

Besides, the man on the run is a deeply American idea. We still rely on the fact that our country is big enough to get lost in. In the back of our minds, there is always an escape route waiting. Such is the promise of our great nation—the promise of escape. It's why our forebears came here in the first place, and why, after they came, they continued fleeing west, in search of wealth, freedom, anonymity. It's the reason that, even today, the road trip remains such a profoundly American tradition—much like the sport of running itself. Because we are always running from something in this country, always extending to ourselves that illusion of freedom afforded by our ample geography. And, in fulfillment of our mythos, we always seem to be coming up with new things to run from.

——

By now we had entered the leafy main drag of downtown Aspen, fussily arrayed beneath the gargantuan indifference of the mountains. The day was blue, the high clouds bright white, their cool shadows unmoving on the forested slopes. After three hours sucking recycled cabin air it had been a shock to walk out of the airport into the pristine expanse of the Rockies—like the feeling you got (quick stride fading to an awestruck shuffle) on entering a vast, high-ceilinged room.

A few minutes later we pulled into the St. Regis, a $350-per-night hotel that, with its venerable brickwork and unimpeachable decor, was clearly striving for an old-world milieu, a milieu in which the super-rich could safely assume that everyone they encountered really did own their own plane.

It made me wonder about James's friendship with Brian, who hadn't a dime to his name. The two of them, in fact, seemed about as opposite as you could get. Brian was from the North, James from the Deep South. James stayed at the St. Regis. Brian had withdrawn to the remote wilderness, an island of pine halfway to the sky and surrounded by desert. They'd only met in person once or twice.

How well Brian and James actually knew each other I couldn't tell. Among their tribe there existed an odd, rootless intimacy that blurred interpretation. It was as if they were all part of the same ongoing story, like they were all reading the same book, allowing them to feel unusually close to each other without really knowing each other at all. And what did knowing even mean, for those for whom so much was unknowable?

But I deceived myself in claiming that this ignorance was theirs alone. For the truth was that I knew no more than they did about the threat posed by chemicals and other noxious substances. I just pretended I knew. Or rather I pretended that I didn't *not* know. Who, then, was I to say that they weren't right when they described themselves as canaries in the coal mine of some impending toxic doom? The suggestion was not totally incredible. And in my own mind, in fact, it accorded with a growing suspicion that the world we'd created for ourselves was probably a lot less safe than we'd like to believe.

In a way, you could say that it was only because we had been so successful in neutralizing all the obvious threats. The really old-school ones like weather, animals, you know . . . bandits. Our world was built as a bulwark against these threats. But now it was starting to seem that this world had itself become the threat.

Certainly toxic chemicals were a cause for concern. But how much of a concern? And how was one to navigate it? After all, the threat was not always so obvious. It didn't roar at you from the tree line, or light up like a nuke. Instead it was woven into the fabric of everyday life. The warnings would come in the form of a small item in your news feed, or a filler story on the radio. Or maybe a friend would let something slip as you stood on the soccer sidelines, watching your innocent five-year-old toddle after the ball. Did you hear that thing about the crayons laced with asbestos? The organic tampons doused with weed killer? Or that popular brand of whiskey tainted by an ingredient commonly found in antifreeze?

Great conversation fodder. Strange enough to be interesting, but never serious enough to warrant alarm. More micro-threats than threats, really.

Still, they kept coming, a steady patter pinging the amygdala. Phthalates in your printer ink. Formaldehyde in your furniture. Glyphosate in your Cheerios. Chlorine in your sex toys. Triclosan in your underwear . . .

Nor could you ever quite get a lock on the danger, for it was always changing, like those periodic updates on the health impacts of coffee, red wine, or jogging, which seemed to reverse polarity every other week. And always obscured behind a fuzzy screen of language. Those toxic scented candles you read about, for instance, could "negatively [impact] your skin and your overall health." Could. Maybe. Somehow. The chemical in

Subway sandwiches was "linked" to asthma. And "potentially linked" to cancer. Potentially. Maybe. You couldn't know for sure.

Even when the really big ones dropped, like research showing that sperm counts among men had fallen more than 50 percent in the last forty years, an outcome attributed to "chemicals in commerce," or that study that found dangerous chemicals in America's favorite kids' food, Kraft mac & cheese, or the report on toxic chemicals in the drinking water of over six million people, at up to twenty-five times the maximum safety level set by the EPA, they were always hedged with that same uncertainty. By the end of the sperm article, for instance, just when you were wondering whether it was time to bid the human race adieu, a scientist would inform you that actually it was "too soon to draw a conclusion." And though the mac & cheese chemical had indeed been "linked" to all sorts of health problems, scientists couldn't say for sure that it *caused* them. Ditto the chemicals in the drinking water.

That was part of it, though. The spookiness. Just never knowing for sure how paranoid to be. And then every so often you'd hear a story that hit a little closer to home. Like the one I heard recently about the son of an acquaintance who was born with hypospadias, a deformity in which the urethra opens not at the tip of the penis but somewhere along its length, and which had been specifically linked to hormone-mimicking phthalates. Or the daughter of another acquaintance who began developing breasts at age eight, and had to take growth hormone inhibitors to delay puberty. Or all the stories about kids I knew with allergy issues, or ADHD, or maybe a touch of autism.

I never used to worry about chemicals. Why would I? I didn't feel sick. I felt just fine. In fact, life seemed to be getting easier and more convenient all the time. Longer, too. One hundred years ago I could expect one more year of life. Today, I could expect thirty-two. Why? Science. Medicine. Chemicals. Decades ago the promise was made: better living through chemistry. And hadn't chemistry delivered?

Even so, it was hard to escape the feeling that the promise was beginning to falter. Probably this was inevitable given how much the human environment had changed during the last 150 years from the one we'd evolved to inhabit. For the first time in history, for instance, we were spending somewhere between 95 and 99 percent of our time indoors.

One day not too long ago I reached a point where my unease finally prevailed over my desperate wish to remain ignorant. Maybe it was becoming a parent. Not just that I was worried about my kids but in having kids I had become more accustomed to hard slogs and to the fact that they could rarely be avoided. At some point you just had to buckle down and deal, like a proper adult.

So I began reading, looking for hard numbers, determined to nail down the threat. Sure enough, there were plenty of numbers available. Like the 85,000 synthetic chemicals in play in our daily environments. Or the seventeen pesticides we were, on average, exposed to every day. Or the nine thousand food additives (over ten times as much as fifty years ago) that thronged the aisles of our supermarkets. All of which eventually funneled into our bodies.

And into the bodies of our children—even before they were born. In 2013, a Canadian organization tested the umbilical cord blood of three newborns and found evidence of 137 different chemicals, including DDT, PCBs, and fire retardants. The finding spoke for itself. Canadian children were born pre-polluted. And that's *Canada*. What would they find in babies born in the U.S.?

Nor did this deluge of chemicals show any sign of abating. In 2017 alone, 3.9 billion pounds of toxic chemicals were dumped into our environment. In the fifteen years prior the chemical industry itself grew from $1.8 trillion to $5.7 trillion. Meanwhile plastics suffused with phthalates, fire retardants, BPA (bisphenol A), polyvinyl chlorides (PVCs), and other petrochemicals continued to gather around us—a total of 8.3 billion metric tons of it as of 2015, over half of which was produced since 2000.

Which leads us to that classic metric of modernity—cancer rates. Thyroid and liver cancer up 300 percent between 1975 and 2014, non-Hodgkin's lymphoma and kidney cancer up 200 percent, and significant increases in acute myeloid leukemia, testicular cancer, and breast cancer.

And that's just for starters. Beyond cancer an entirely new class of threats seemed to be emerging from the twenty-first-century closet. Autism, for instance (up 220 percent between 2000 and 2017, with environmental toxins suspected as likely contributors), mental retardation or intellectual impairment (up 63 percent among children between 2001 and 2011), allergies (sensitization rates approaching 50 percent among children), obesity (highest prevalence in recorded history in the U.S., and convincingly linked to chemical exposure), genital birth defects (increasing), and what one recent study called an "epidemic of male reproductive problems."

And sure, much of what I'm saying here is the same sort of thing you might find in a number of overwrought books about chemicals and the environment. But I swear I'm trying to be clear about it. I am trying to be sane, as sane as one can reasonably expect to be when so little is known for sure. I am not, for instance, piling on cancer stats that could easily be explained by improved diagnostic methods or classificatory changes. I have not cited trends in Alzheimer's or Parkinson's, because despite the increased incidence of these diseases that has come with an aging population, the prevalence actually seems to be decreasing. And while some scientists do indeed suspect that chemicals may play a role, as yet no one has managed to demonstrate even a compelling correlation. For the same reason, I've said nothing about asthma, the prevalence of which, while increasing, is not on the same order as autism.

And to be clear, to be sane, even the studies cited above invariably conclude on a familiar note of uncertainty. "The problem," the author of the thyroid study remarked, "is that PBDEs [fire retardants] are very widespread, so proving a link between exposure to the chemicals and an increase in thyroid cancer is a real challenge. There is just very limited data."

Or, to quote the author of the autism study, "only application of a comprehensive set of criteria for assessing causation combined with a deeper understanding of the underlying biology and epidemiological evidence correlating individual-level exposures and outcomes can prove whether a suspect compound or trigger is a likely cause."

Which was basically just a long-winded way of saying: Only God could know for sure. And, as such, the true magnitude of the threat remained anyone's guess.

In fact, so little was known about the effects of toxins that many were describing the situation as "a vast human experiment." Conducting experiments on humans has been illegal in the United States since the passage of the National Research Act in 1974. But considering that we're talking about a society that permits the introduction of hundreds of untested new chemicals every year, despite the absence of "a comprehensive set of criteria for assessing causation," limited "understanding of the underlying biology," and "epidemiological evidence" that was sketchy at best, what the hell else could you call it?

Uncertainty always has a corrosive effect, raising new doubts about old dogma. To what extent could we really say that we even knew ourselves, for instance, when our most primitive metric of self-knowledge (our vulnerability to external threats) was so clearly obsolete? In simpler times it was enough to know that the human body worked best when not perforated by a spear. But now it seemed our vulnerabilities were far more extensive and profound. The new truth was that the human body was basically porous. And this in turn raised serious doubts about our competence as caretakers of our own flesh. I mean, how do you shield yourself when the spears are everywhere?

It made me feel like a kid again, and not in a good way. When I was six, I'd hold my breath passing cemeteries. Because why risk it? Forty years later I often found myself doing the exact same thing. Holding my breath when a diesel truck rumbled past, or for those first few seconds after starting the car when the vents blew odd-smelling air. Or on venturing into my basement, and the faint paint fumes that lingered there, because I thought I remembered reading somewhere that paint fumes were bad news.

It made me wish that I could turn (like a kid) to some trustworthy authority for advice and reassurance. But there was no such authority. The only entity capable of playing such a role was the government, and the very idea of a vigilant, paternalistic government had long since become laughable. In fact, in the United States chemicals were generally assumed to be safe until proven otherwise. The vast majority of the 85,000+

chemicals in our environment had never been tested either individually or in combination for their effects on humans. Not even children's toys were required to be independently tested before entering the marketplace. "Responsible companies test their products before offering them for sale," the Consumer Product Safety Commission (CPSC) assured us. Because all companies were totally responsible, right?

The truth is, the Food and Drug Administration, EPA, CPSC, and other such regulatory bodies were no more equipped to monitor such a vast and complicated array of chemicals than the Securities and Exchange Commission was to monitor Wall Street. And the enormous amount of money that the chemical industry slathered on Congress ($100,000 per member, on average, in 2016) ensured that this wouldn't change. In the summer of 2016, in fact, when I began investigating EI, we were on the verge of electing a president who had explicitly pledged to "put the regulations industry out of work and out of business." A pledge on which he would soon deliver, removing a ban on lobbyists working for government agencies that they had lobbied within the past two years; appointing a known climate change denier and close ally of the fossil fuel industry to head the EPA; and appointing an FDA chief who had previously served as the director of eight different pharmaceutical companies.

What was the average citizen to do? Soldier on and hope for the best. Cobble together some ramshackle list of precautions based on anecdotal knowledge, sporadic news reports, and whatever your budget could afford. I could afford to buy organic, so I did that, despite studies suggesting that doing so could actually make you less healthy. I washed my hands frequently, threw out my Teflon pans, and tried to remember not to microwave my Tupperware.

But I did these things less with confidence than a kind of helpless shrug. After all, if it was impossible to gauge the danger then it was equally impossible to know which precautions were necessary, which were needless, and which were no better than a prayer. The vast, invisible risk calculus I'd been carefully nurturing since I first stumbled as a child suddenly no longer applied. What, for instance, was the risk of canned tomato sauce (lined with BPA plastic) versus driving without a seat belt? What was the risk of that new car smell (the product of 275 different volatile organic compounds) versus four glasses of wine or a weekend hang-gliding course?

There was no way to know. A certain level of tacit helplessness had become the default, requisite to participate in society at all. Ultimately you could only accept that, like folding maps and fresh air, like glass ketchup bottles and fish free of mercury, a certain caliber of certainty was simply no longer on offer. Instead of mercury-free fish you had a 25 percent increase in books on anxiety. Instead of fresh air you had a new Canadian company selling bottled air from Banff.

This diminution of certainty was partly why I became so interested in Brian. Because even though sensitives were often dismissed as crazy, they were the only ones ready to acknowledge that we really were living in the midst of a vast human experiment. It was as if their role as pariahs had allowed them to say something that the rest of us were no longer able to say, or had forgotten how to.

Here's where EI differed from all those other mysterious illnesses, like fibromyalgia and ME (myalgic encephalomyelitis). It's not just that it was underreported, and impacted four times as many people. It was that those suffering from EI attributed their suffering to a specific cause. A cause which happened to include everything that made modern society what it was. Carpets and electronics. Mattresses and car exhaust. It wasn't them that needed curing, in other words. It was modernity itself. Thus some referred to EI as the "Twentieth-Century Disease"—now the Twenty-First—the flip side of our profligate consumerism and the ever-expanding shitstorm of chemical detritus that made it possible.

But it wasn't only this that made me come to think of EI as the defining ailment of our time. After all, EI was not the first disease to implicate modernity. Cancer had done the same thing several generations earlier. What made EI the defining ailment of our time was how hard it was to know if it was real.

CHAPTER 3

The Comfort of Guns

James was feeling kind of bad now, he said. Not as bad as last week, when his family was here. Since then, he'd done a few sessions in a hyperbaric chamber, which somehow lessened his symptoms. These he described as fatigue, muscle ache and general inflammation, shortness of breath, and a feeling like his throat was closing up. But worst of all was the head pain.

"It's like . . . at random times during the day there's like . . . a very sharp icepick drilling . . . into my head."

He spoke haltingly, gullet working, as if he were trying to swallow a belch. We were sitting at a wooden table at the edge of a stone-tiled courtyard behind the St. Regis, with a clear view of the mountain's grassy ski slopes. Seated again, James had regained some of his poise, becoming once more the eligible Aspen bachelor. His polo shirt revealed a fit chest and sizable triceps. His large, rectangular fingers handled the menu with confidence. But when he groped for words to describe his pain, his face and right hand contorted, dispelling the illusion.

It's always hard to know how to read another person's pain. Nothing feels more real than pain, yet nothing sounds more remote. In the online EI support groups, Brian was always talking about it. "Constant pain," "satanic excruciating pain," "crazy pain," "pain so aggressive," "intense nervous system pain," "pain from exposure and nervous system upregulation," "physical, mental, and emotional pain," "paralyzing pain," "pain the regular world has yet to fully realize exists." For him it seemed a way of insisting on the reality of his subjective experience. But the language in which pain is written is always private, and virtually impossible to translate. Such dictionaries as are available are usually comprised of the most embarrassingly primitive vocabulary, a range of cartoon faces, say, of the sort you might find tacked to a kindergarten wall. Mostly, we accept the pain of others on faith.

In credentialing this pain a medical diagnosis would have spared me the labor of faith—or else the crummy feeling that came when faith failed. In this light, a diagnosis seemed not unlike a social custom, evolved to help the hale feel more compassion and the sick feel less alone.

Even without a diagnosis the language used to describe James's condition could still color how he was perceived—and this in turn could have real impacts on his life, perhaps even influencing his eligibility for worker's compensation or health insurance benefits. I'd heard it said, for instance, that the chemical industry favored the term "Idiopathic Environmental Intolerance" because it did not include the word "chemical." By the same token, environmental medicine doctors and sensitives themselves preferred the term "chemical injury," which clearly stipulated cause and victim.

Without a clear diagnosis the language used to describe EI was up for grabs. And the only hope for a clear diagnosis was in the biology. Not that it was necessary to know everything about how EI worked. But at the very least what was needed was a reliable biomarker—and a way to test for it. But there was no such thing. Folks had been searching for years but so far they had found nothing.

Well, okay, not nothing. In fact, dozens of biomarkers had been identified. Changes in heart rate, for instance, or decreased blood flow to certain areas of the brain. Noise sensitivity or problems with balance.

exposure experienced by veterans remained unknown. And anyway the animal studies still failed to identify a mechanism, or explain why some veterans exhibited multiple symptoms and others had none, or why GWI struck veterans who had never even deployed to the Persian Gulf. Which is why the Academy of Medicine concluded their 275-page report—the last of ten—by essentially dismissing GWI as a product of "deployment related stress" and encouraging researchers to move on.

James's headache did have one advantage: unlike Gulf War Illness symptoms, and other common EI symptoms like fatigue and brain fog, an icepick headache was difficult to dismiss as purely psychogenic in origin. Like EI, headaches, too, lacked reliable biomarkers, but in the last thirty years a gathering body of evidence had shown, for instance, the role of the hypothalamus in cluster headaches, the role of the brain stem in migraines, and the role of the aptly named MTHFR gene in migraines with aura, and with this the psychogenic explanation had taken a backseat.

Recent decades, in fact, had seen a huge explosion in pain studies, a phenomenon mainly attributable to ever more powerful imaging technology and a ballooning interest in chronic pain, which affected over 25 million Americans. The growing body of research not only challenged long-held views on the distinction between mind and body but also contributed to a gradual shift in the way patient reports of pain were interpreted—particularly in the wake of the opioid crisis, which, in furnishing a proper villain (pharmaceutical companies), finally allowed chronic pain patients to be recognized as victims.

Consequently, what was once dismissed as "psychosomatic" (that is, made-up) pain was now generously referred to as "centralized" (originating from the central nervous system—as opposed to some clearly identifiable point on the periphery). "Somatoform disorder"—the *DSM-IV* diagnosis that included chronic pain, as well as everything from body image problems to hypochondriases—had itself been redefined to reflect that medically unexplained symptoms (MUS) could no longer be automatically categorized as merely psychological.

And finally, researchers were beginning to notice that women experienced pain differently than men. Finally the assumption that one body

Neuro-ophthalmological symptoms. Autonomic nervous system
liarities. Changes in cytokine behavior or antibody response. Evei
creased dermal sensitivity to hot peppers.

So there were indeed reasons to think that what sensitives we
ing was in some way "real." But while these reasons were found
sensitives, none were found in *all* sensitives. They wouldn't nec
have to be if they all derived from a common underlying source. E
could such a diverse range of biomarkers possibly have in comm
was it even clear that sensitives' sense of smell was any more
than anyone else's. And that's what was really wanted. Something
up their story line, a definitive biological sigil that proved they w
they said they were, dread heralds whose near-biblical burden wa!
tell the coming chemical cataclysm—or at least confirm it witl
authority than the rest of us.

The most promising EI-related research came from studies
War Illness, which presented similar symptoms and was pres
have something to do with the many substances (chemical warfai
chemical warfare antidotes, vaccines, insect repellents, smoke fror
fires, etc.) to which soldiers were exposed while deployed. Betw
and 2014 federal funding for research on the health of Gulf War
totaled more than $500 million. And it did yield some results. For
the brains of soldiers with GWI had been shown to exhibit struc
ferences, which in turn were associated with increased fatigue,
cognitive problems. And studies of animal models had proved
tiple toxins administered in combination could produce the sar
toms seen in GWI—and at levels that would have produced no r
they had been administered individually.

But even the GWI research was deemed inconclusive. As a
Academy of Medicine report spelled out, none of the research
had been able to pinpoint a disease-specific biomarker, nor ev
marker to indicate exposure to a particular toxin. Nor was
way to be sure that the biological irregularities identified by tl
weren't simply due to the stress of combat. As for studies involvi
models, their relevance was limited by the fact that the actual

could stand for all—like da Vinci's drawing of the Vitruvian man, the nude dude in the snow angel pose—was giving way to a more complicated view. There was not one kind of human body. There were at least two, and probably many more. After all it wasn't just pain that people experienced differently, but pleasure, too. And pharmaceuticals. And sitting still for long periods of time. And food. And a host of other things. How many models were there really? Two? Two hundred? The answer was unknown and likely unknowable.

Thus, while the medical community remained dogmatically skeptical of EI, in other areas room was slowly being made for a certain level of uncertainty. It was as if the view of human beings was finally catching up with the way atoms were viewed ninety years ago, when Heisenberg introduced uncertainty as a permanent and necessary feature. Whether the delay was due to a lag in technology or simply the distracting compulsion in medicine (absent in pure science) to maintain authority over its human subjects, patients were now being accorded greater credibility—a positive development, with maybe even far-reaching implications given how evil always seemed to commence with a denial of other people's pain. It suggested a gradual broadening of the custom that allowed the hale to feel more compassion and the sick to feel less alone, as well as a future in which industry would have less power to dictate the reality of those impacted by it, and doctors would have less power to dictate the reality of their patients—particularly women, whom EI affected in greater numbers, and whom traditionally patriarchal fields like medicine had neglected and belittled from the dawn of time.

To the extent that operative truths boiled down to credibility, then, perhaps the scientific basis of EI mattered less than I supposed. Maybe all that really mattered was that James ended up looking marginally more credible than the chemical companies. Bearing in mind that quip about how corporations, if they really were people (as the Supreme Court would have us believe), would probably be regarded as sociopaths, plus the well-documented villainy of the tobacco industry, whose playbook the chemical industry zealously followed, this seemed reasonable to expect.

Of course, there would always be an inhuman predictability to corporations that made them perversely appealing. Unlike James, for instance, you wouldn't expect a corporation to randomly pull out a gun.

———

"Glock 40," James said, popping the magazine to flash the payload.

We had moved upstairs to James's hotel room so he could start packing his stuff and get on the road. The gun had emerged casually from his kit bag. Handsomely sinister, raven black, it looked misplaced against the elegant taupe decor, like it belonged to some other story line.

There is a certain mesmerizing density to a gun. It is as if absolutely everything extraneous to its cold purpose has been leached out, leaving only this dark snarl of lethality.

How do you respond to a drawn gun? Was he subconsciously sending me some kind of message? There was no obvious menace to James. On the contrary, the more we talked the more harmless he seemed—boyishly enthusiastic, painstakingly polite. When he laughed he thrust his head forward, as if we were sharing a secret joke.

But this apparent harmlessness was in its own way worrying. He seemed almost too anxious to please, like an intern on his first day—an impression which didn't quite square with my image of him as a Southern scion and successful entrepreneur. A puzzle piece was still missing, or perhaps several, and it would be hard to relax until I knew what they were—particularly when he was still gripping the gun.

Then again, it's not like he didn't have a good reason to carry. A world of enhanced uncertainty called for enhanced precautions. James cited this one time in Kayenta, Arizona, pulling into a gas station late one night and being approached by a girl asking for money. And as he was talking to her he noticed these two other guys circling around behind him. It was one of those times when you couldn't totally be sure if you were in danger, but it would have been nice to have a gun, just in case. Guns were not unlike ibuprofen in this way. They made you feel better.

Besides, James said, with the detox regimen he was on he was always pulling over by the side of the road to pee. And you never knew what might pop out of the bushes. Once he stopped just outside Phoenix and there was this sign—*Beware Hitchhikers, Prison Nearby*. And where does something like that fit into the risk calculus? With a gun, you didn't need to know. With a gun, you could pee wherever the hell you liked.

Then there was that time at his house in Sedona, two years ago. He was sitting on his back patio wearing only his boxers, eating a breakfast of tofu and greens in coconut oil and soaking up some vitamin D. The patio overlooked a small backyard with a dramatic view of the Sedona hills. It was a good day, one of those days when James actually felt close to normal. And then out of nowhere this mountain lion appeared.

"And it was just ripped," James said. "I was amazed. I was, like, 'This thing is beautiful.' It stopped and looked right at me. And I've read or heard that you don't look a mountain lion in the eyes, but I did. I was thinking, 'Fuck, I'm going to have to fight this mountain lion in my underwear!'"

So, sure. A gun at that moment, laid out like any other utensil on the breakfast table, might have been welcome.

———

James reholstered the Glock and began gathering up pill bottles, jumbling them into a black foam-lined suitcase. Opening dresser drawers, he removed an oxygen regulator, a heart rate monitor, a temperature probe, a blood pressure gauge. The heart rate monitor, he said, sync'd to an app on his phone so he could give his doctor live updates.

"I've got Mānuka honey, too," he said. "Some people believe it has healing properties."

I watched, wondering. He was an odd hybrid. Affable, credulous, unwrinkled by irony, the product of what increasingly seemed like a conservative Southern upbringing. Past the crew team exterior you could almost picture him ambling down a country road, hayseed between his teeth, without a worry in the world.

Yet somehow he'd crossed over to the opposite cultural shore, where the norms were not guns and Republicanism but veganism and yoga. Southern California, where we were ultimately bound, was where he felt most at home, he'd said.

I had hoped that his economic background would shed some light on his character. On the one hand inherited wealth could have argued against him, in neutralizing the credibility he might have gained by achieving financial success on his own. But it also might have helped him, in contextualizing his eccentricity. Eccentricity is, after all, the natural right of the wealthy in this country. Rich eccentrics embody the dark flowering of the American ideal, signifying that magical point at which reality itself can be bought.

As it turned out, James's great-great-grand-uncle was a figure of some importance in the small Southern town where James grew up. He gave a ton of money to build a children's hospital—the same hospital where James used to get his allergy shots. He also founded a local bank, which today, James said, remained his best lender.

But James was quick to dismiss any notion that he had benefited from dynastic wealth. None of his great-great-grand-uncle's money trickled down to his side of the family. His father was a self-made man.

"He put me to work early, y'know?" James said. "He's, like, 'Oh, you want this? All right. Go work.' So I planted trees and did a lot of yard work. I planted thousands of trees when I was younger."

James's father was into tree farming at the time. When he got older James went into business with him, along with his brother, but split off a few years later to pursue his own vision. He was always the rebel of the family, he said, a skate punk during his high school years, and, after that, one of the few kids who didn't follow the crowd to the state university. Instead, James opted for Stetson, a private university in central Florida. Later he transferred to American National College in Atlanta—mostly because he had gotten into flipping cars, and Atlanta offered a larger market.

James had acquired his first car dealer's license two years earlier, at age seventeen. It was his father's suggestion. His father was always pushing him in business.

"I learned a lot in my dad's office," James explained. "When I was very young, like, elementary school, in the afternoons . . . I would walk to his office and he would give me something to do. Like . . . 'I want you to go to the county and file this deed.'"

It was an odd image, an elementary school kid filing deeds for his father in some sleepy Southern town. Why would any kid that age choose to spend his time filing when he could be out horsing around with his pals? And how could James claim to have been a rebel when he was always working for his dad?

After a certain age, most people instantly make sense to you. Usually you've seen some version of them before. Their choices and failings are familiar, as are the long list of compromises that has delivered them to their current station. You learn to recognize a certain pattern of pretense — who they are pretending to be, if not who they are.

But I didn't get the feeling that James was pretending to be anyone. At heart he still seemed like that blue-eyed country boy, game and gangly, canny enough to succeed in business (or else just preternaturally lucky, the way innocents can sometimes be) but entirely devoid of the usual adult subterfuge. As far as I could tell all he really wanted was for me to like him, and for us to get along.

Maybe the usual posturing and pretense of social identity had ceased mattering to him much as it had ceased mattering to Brian. Or maybe the confusion I felt said less about him and more about me. I liked to think of myself as kinder and more thoughtful than EI's cynical critics but in some ways I was just as cynical, if not more so. Perhaps this was part of what inspired my search for Brian in the first place: self-disgust. A desire to be free of that miserable cynicism. To "off-gas" it, as the sensitives said. Maybe it was possible. Or maybe the only way to truly escape the cynicism was to become sick yourself.

CHAPTER 4
Crazy Charlie

I suppose it was lucky that James opted to stop at the bike shop after leaving the St. Regis. Sometimes you need to see how someone interacts with others to understand where they're coming from. And you couldn't have asked for a better foil than Charlie.

"There he is," James said. "What are you up to?"

"Working," Charlie said, barely acknowledging him. "You ever seen it before? I saw a TV show one time, they had pictures of people working. And then they had something else. Oh, it was orange juice, that was it. What's up in your world?"

Around James's age, or maybe a bit older, smiling myopically behind thick glasses, Charlie wore a pink-and-white-striped button-down shirt outside of cargo shorts and spoke in a semicoherent bluster, with whole words smeared and elided in the tumble of his sentences.

Charlie was the bike shop's owner. James claimed to be friends with him, having met him twelve years ago when James bought his first house in Aspen—one of "four or five" that he'd owned over the years.

But the vibe coming off Charlie was sneering, bordering on hostile.

Most of his talk took the form of bullying, byzantine wordplay punctuated by self-admiring snorts. It was as if he were playing to an adoring audience that only he could see. James gamely tried to play along, but it wasn't pleasant to watch.

"Wait, so what are we going to do about our deal?" Charlie said, finally expressing some real interest in the conversation.

"Whatever you want," James replied.

"No, not whatever I want. What I want is you give me your car. And then you take a knife and put it in your ear. Four inches."

He gave another glottal snort and went on, "So when does it need to be funded? When do you want money?"

"First we're gonna close the deal," James said, seeking some sort of anchor. "Second, I'm gonna make sure everything is good with you and Rob—"

"What if Rob has, like, toenail fungus," Charlie interrupted. "That wouldn't be good. You said everything's good with him, right?"

Somehow, through all of this, the deal got discussed. It appeared to be some kind of property deal. James would form a limited partnership and offer Charlie discounted shares. Once Charlie had ascertained what he needed he walked out of the shop midconversation. There, he was immediately hailed by a few locals and it became clear that he was something of a fixture in the community. The eccentric, abusive bike shop owner who imagined himself to be Falstaff. Why James would go out of his way to help such a creature was a mystery.

James followed Charlie outside and stood around awkwardly as Charlie chatted up the locals.

"I'm out of here, Charles," he finally ventured. "Save me a couple bikes."

"You want to buy a bike today?" Charlie said over his shoulder. "No. Okay, go away."

"You have a Thule rack?" James persisted.

"No."

"All right," said James. "Well, I'll come back with one and then I'll get the bikes and hopefully find a place to stay."

"This conversation has no value," Charlie said. "Will we have the same conversation two weeks from now when you come back?"

"We could," James said.

"No. Even if we did, would we. We *could* do anything. We could cut my leg off. We could. We're not going to."

"Maybe we should, though," I said, unable to resist.

Finally, we made it to James's Range Rover. A tricky silence prevailed as I waited to see if James would offer some explanation or apology for Charlie—or why anyone would choose to put up with him. It wasn't forthcoming, and finally I had to ask.

"He's just like that, I dunno," James said lightly. "He's a good guy. You just gotta get to know him."

I let that sit for a bit. This was the first really odd behavior I'd seen from James. The Glock I could understand, in theory. Even his various contradictions might yet resolve and make sense on some level. But what could possibly explain why someone would be willing to put up with such abuse?

I hadn't known James long enough to pursue the question, and so let it drop for the time being, and in affable silence James piloted the Rover back out of town. Gradually the land opened up and there came a quiet sense of relief as our own silence merged with the wider silence of the sky and the hills, the afternoon light falling slantwise on the pleated slopes. A fingerflick of rain dappled the windshield and James pointed to where he once saw a bald eagle scoop a trout from a river. Somewhere around here, he said, Doc Holliday succumbed to tuberculosis.

It was one or two day's drive to where Brian had disappeared, in northern Arizona, but I hoped to visit Snowflake en route. The town of Snowflake (its name taken from Erastus Snow and William Flake, the pioneers who founded it) was something like the capital of the EI world—or at least its most notorious landmark. Located in the baking scrublands of eastern Arizona, Snowflake had become a haven for many of the more severely afflicted sensitives, who, shunned by their families and dismissed by the medical establishment, could live openly there without fear of judgment.

For this reason some of the more conspicuous EI behavior was on display in Snowflake—papering the walls with aluminum foil, hanging mail from clotheslines, storing cell phones three feet underground—at least according to a few gawping news articles I'd read. To me these articles seemed as much a commentary on their audience as their subject. For behind the voyeurism one could sense (as with every such article about the cultural fringe, e.g., preppers) a telling anxiety about whether the weirdos might be right.

But this did nothing to change the way the articles were perceived by their subjects—as yet another insult. One article in particular, in which a *Guardian* reporter manipulated the Snowflakers to gain their trust, left many sensitives feeling profoundly betrayed. "Think of [the media] as vampires," one sensitive wrote. "Once invited, the terror begins."

With this, outsiders' access to Snowflake had come to an end. There was the feeling of wagons being circled, and a fierce privacy descending. We therefore had no idea whether we'd be allowed to visit Snowflake. Nor whether doing so was even safe. Given the way they'd been treated by the media—not to mention the medical establishment, and often their own families—there was no way to know the depth of hurt and fury they felt, or how they'd respond in person.

Today, at any rate, we'd be going no farther than Basalt, an overflow community around twenty miles outside Aspen. The St. Regis, for all its pomp and grandeur, was not agreeing with James, but Basalt had a hotel—the Element—that he had tolerated well in the past. James sometimes had to switch rooms as many as five times before finding one that worked for him. His excuse was always that he had severe allergies, because experience had taught him that the hotel staff could tolerate the truth no better than he could tolerate their rooms.

The Element was conveniently located right across the parking lot from a Whole Foods, and over dinner in their cafeteria James told me stories about how he used to lift weights and work out with semipro athletes.

"I used to be a lot bigger," he said brightly. "Like insanely strong. I could do 365 twelve times. I was a monster."

I was starting to feel exhausted from trying to reconcile all the different sides of his character. Later that night I lay in bed listening to the petroleum streak of cars on the highway, reviewing the puzzle pieces. The tentative way he walked, as if barefoot on hot pavement. His deferrals in traffic and at doors. The way he hunched his shoulders and giggled, like an old lady pretending to be naughty. Then the Glock. The entrepreneurial success. The physical strength.

"A monster," he'd said.

Then there was Charlie. Unable to sleep, I got out of bed to find out what I could about him. The first thing I discovered was that I wasn't the first to be put off by him. Yelp was filled with one-star reviews. "Absolutely appalling," one customer wrote. "Possibly raised by wolves."

Further research revealed a possible cause for his eccentricity. Fourteen years ago, two years before James met him, Charlie was trying to set a new mountain biking speed record on a Snowmass ski slope when he flew over the handlebars and was knocked unconscious, coming to rest two hundred feet farther down the mountain. Observers reported he was going over 90 miles per hour at the time. When he emerged from the coma several days later he could no longer see straight and his speech was slurred. It made me wonder whether the head injury might also account for his lack of social graces.

"It's called a filter," Charlie had said. "I don't have one."

Would a medical diagnosis have made Charlie any more sympathetic? Who knows? Maybe he was always an asshole, and just lost the ability to pretend that he wasn't.

———

Theorizing was a favorite pastime in the land of EI. Everyone had their own way to explain the disease. Some offered a neurological account; others were more psychological, or biochemical.

The theories had evolved over time. The original explanation for EI emerged from an allergy model, in which symptoms were mediated by the immune system. In this scenario, chemicals behaved a lot like pollen, or any other common antigen, somehow triggering an inflammatory immune response despite being many times smaller.

As more became known about how the brain worked, the allergy model shifted to a more neurological footing. In 1991, Iris Bell, a psychiatrist at the University of Arizona Health Sciences Center, combined elements of biological psychiatry and neurotoxicology to develop a theory that came to be known as neural sensitization, which suggested that EI was a consequence of the open pathway between the nose and the limbic system—our lizard brain—which allowed chemicals to bypass the body's security protocols and wreak havoc with our most primitive neurological functions. Repeated low exposures to chemicals, the theory went, could sensitize the limbic system in a process known in biology as "limbic kindling."

Bell did not exactly spell out how a sensitized limbic system could lead to the highly diverse symptoms that sensitives experienced, but this was the beauty of a neurological theory: because the limbic system was connected to everything (the autonomic nervous system, the endocrine system, the immune system, etc.), pretty much any disturbance in bodily function could in some way be traced back to it.

Two years later, William Meggs, a doctor of emergency medicine and aspiring clinical toxicologist at East Carolina University, introduced a complementary theory suggesting that EI might be attributable to "neurogenic inflammation," a phenomenon first discovered in 1910 by a Scottish doctor upon observing that mustard oil squirted into the eyes of rabbits provoked tissue inflammation, but that inflammation did not occur when the nerves were blocked with a local anesthetic. According to Meggs, this proved that tissue inflammation could be triggered not just by an immune response to antigens like pollen but also by irritating nerves. As in Bell's account, the fact that the inflammation was mediated by the nervous system explained why symptoms could crop up anywhere, via "aberrant or conditioned crossing of [nerve] pathways." Or maybe both.

The neurological theories had several advantages over the allergy model. Not only did they explain the wild array of symptoms sensitives experienced; they also offered a sensitization process which might explain why sensitives responded so dramatically to toxin levels that barely registered with anyone else, as well as why the list of chemicals they could not tolerate always seemed to grow longer—a phenomenon known as "spreading." It could all be attributed to the central nervous system: the more sensitized the central nervous system became, the more violently it responded, and the more agents it perceived as threats.

What was lacking in these theories—and a host of other, less convincing theories—was a way to test them, in part because they were all so speculative and vague. The problem seemed to be endemic on the frontier of the unknown: the more you looked for answers the more questions you encountered, such that an inquiry that began by asking "What's making these people sick?" soon led to much stickier questions, like "How exactly does the nervous system channel inflammatory signals from one part of the body to another?" and "How does the human body work?"

Eventually it left you feeling that, however deeply you looked, there would always be something deeper, a level of magnification beyond which you could not peer. It was one thing to recognize this of atoms or quarks, whose extremities of scale assured them a relevance comparable to that of distant stars. It was something else to recognize it of one's own body and understand that the mystery of that body—how the minuscule mishaps of 37 trillion cells somehow translated into sadness or fatigue—might always be beyond your reach, no less opaque, in the end, than the body of an alien. Indeed, if anything emerged from reading through so many theories it was a sense of how, in the aggregate, they reflected the human body's insurmountable complexity. Perhaps EI had simply learned to capitalize on this complexity. Or else in searching for answers researchers simply saw what was there. That is, everything.

In this respect, there should have been something deeply satisfying about the work of Martin Pall, a widely published professor of biochemistry at Washington State University. In 2000, Pall released a paper detailing a biochemical explanation for EI. The paper was unprecedented in its specificity, mapping out a complex biochemical cycle (involving five smaller, interrelated cycles and over twenty-two distinct mechanisms) to explain EI's root cause.

For Pall, the problem began inside individual cells and how they responded to foreign threats like chemical irritants. Where normal cells reverted to their usual placid selves once the threat was neutralized, EI cells,

Pall thought, got caught in a vicious biochemical cycle, unable to switch off. Pall called it the NO/ONOO-cycle in honor of the elevated levels of nitric oxide and peroxynitrite that were central to its function.

Pall's theory was compatible with Bell's and Meggs's theories, and thus able (like them) to account both for the "spreading" phenomenon and why sensitives were so sensitive. The vicious cycle it described accounted for the chronic aspect of EI (once the cycle got going it wasn't easy to shut down). And the complexity of the cycle afforded a variety of different points whereby external toxins could enter the cycle and disrupt it. The biochemical vocabulary of Pall's explanation, in fact, was so elementary that he saw no reason why it couldn't also be used to explain an entire cohort of mysterious ailments, from myalgic encephalomyelitis to fibromyalgia.

The problem was that, unless you were a biochemist, Pall's theory was all but incomprehensible. The diagram of it looked like something from the playbook of a deranged football coach, with arrows pointing every which way, looping, intersecting—less like a diagram than a Gordian knot.

Perhaps it was foolish to expect that the answer would, if discovered, immediately make sense. Like the expectation that, supposing a lion could talk (per Wittgenstein), we would understand what it had to say. There always had to be some intermediary anthropomorphism to render alien meanings for the human perspective. That this perspective was least in evidence in the workings of the body's own biochemistry only highlighted the gulf between the mind and its biochemical origins.

Pall himself had speculated that the relative crypticness of the NO/ONOO-cycle was part of why it was so rarely cited in the literature. Far easier to stick with psychological explanations that were more native to the human frame of reference. Or even semipsychological explanations like "behavioral conditioning." In this view, repeated exposures led sensitives to begin associating certain odors with certain feelings or experiences. It was the same theory that explained why high-end hotels and shopping malls misted their lobbies with perfume.

The advantage of the conditioning theory was that you could leave the biochemistry out altogether. It wasn't about cells. It was about behavior. In fact, the further you got from the biochemistry the more easily EI could be framed as a character flaw, a psychological anomaly in which the onus of the illness rested firmly with the patient. Pall's explanation, by contrast, narrowed the focus to cellular mechanisms shared by everyone. Which made it that much harder to ascribe the disease to a particular patient's identity. How the blame was framed made all the difference.

DAY
2

CHAPTER 5

Ignorance Ain't Easy

"See, 8," James said, noting the pollen count on his Kestrel, a palm-sized device that tracked a range of environmental metrics and connected to his phone via Bluetooth. "To me that explains why I'm kinda feeling like I need my Claritin D and I need my Flonase."

We were outside the Element, sitting on the Rover's tailgate, where so many of our impromptu conferences would take place. The day was windless, the air dry and clean. Eighteen-wheelers gurgled past on Route 82.

It had been a weird morning. It began with James's suggestion that we stock up on provisions at Whole Foods: coconut chips, a few more bottles of high-pH Fiji Water, and a handful of "paleo-friendly" lamb bars, which were packaged like protein bars but made of 100 percent organic grass-fed lamb meat. This was another of his discoveries.

"To me it's a lot better than lamb jerky," he said.

Which was interesting because I'd spoken earlier to another sensitive who also favored lamb. Lamb, in fact, was one of the two foods he could eat, the other being lettuce. I mentioned this to James and he responded with enthusiasm. One of the booklets his doctor had given him had said that sensitives did well with lamb.

Now James was walking me through his daily environmental analysis. According to his phone, the threats this morning were chenopods, grasses, and sagebrush. What the hell were chenopods? I wondered. They sounded sinister.

"I don't know," James said. "If it was an 8 and it said ragweed or juniper, I would have already checked out by now and been outta here, quick. I would take a bunch of Claritin, throw on the mask, and go."

The humidity was decent, anyway, I observed.

"It's under 55. We don't know what the mold spore counts are, really, but I know mold incubates at 55 percent humidity, so."

Mold was one of James's bête noires. It was a mold exposure that had originally triggered his EI in 2013.

Much of the EI community was obsessed with mold. For many it seemed to represent a scourge worse than nuclear weapons, the internal combustion engine, and plastic shopping bags all rolled into one. The magazine articles that regularly appeared claiming to debunk the mold threat only further stoked their zealotry.

"I have been predicting for years that my situation is not unique, and an epidemic of mold reactivity is certain to occur," one Old Testament "moldie" proclaimed on Facebook. "That epidemic is now happening."

Brian, too, had cited mold as a contributing factor. He'd been coping with low-level symptoms for several years before being advised by his uncle, a general contractor, to remove the carpet in his basement. But the carpet was just a lid on a seething mantle of mold.

"That's when I really went into this death spiral," he'd said.

On the face of it, mold made for a rather disappointing villain. And the online jeremiads against it invariably struck a faintly comical note, like a Holy Crusade against ink stains.

There were different kinds of mold, of course. And sensitives were actually less concerned with mold per se than mycotoxins: the toxic chemicals secreted by mold. Exposure to mycotoxins usually resulted from ingesting grains, which often served as a host for mold growth. And grain contamination was common enough that mycotoxins could probably be detected in many people—in negligible amounts.

Which isn't to say that mycotoxins weren't a real threat. There were several cases in which ingesting mycotoxins had proven fatal. During World War II, for instance, starving Russians had been forced to eat grain left in the field. Thousands were sickened, entire villages wiped out. In fact, the International Agency for Research on Cancer had classified several mycotoxins as carcinogenic, and several others as "possibly" carcinogenic.

Whether inhaling (as opposed to ingesting) mycotoxins could lead to negative health impacts, as sensitives liked to claim, remained an open question. The Institute of Medicine dismissed the possibility, citing "insufficient evidence," a finding that was echoed by the World Health Organization.

But this could be read in several ways. Just because sufficient evidence had not been found didn't mean it didn't exist. In the field of mycology much remained a mystery. To date, for instance, 100,000 different fungi had been catalogued. It sounds like a lot. But it had been estimated that ten times that number had yet to be discovered.

In this, the state of knowledge about mold perfectly paralleled the state of knowledge about chemicals. What's more, mold, like chemicals, could potentially be anywhere, and thus could turn an object as seemingly innocent as a clothes iron (with its inner reservoir of water) into a potential source of evil. Which is why James always packed his own iron.

It put me in the mind of something Elie Wiesel once said about the polluting effect the Nazis had on language, rendering once-wholesome words like "camp" and "oven" unusable. The net effect was as if all the innocence had been drained out of the world—an effect which made James's boyish enthusiasm all the more conspicuous.

Then again James's innocence might well have been what let him remain relatively engaged with the world despite facing so much uncertainty. This was another difference between James and Brian—how they responded to uncertainty. Both gravitated toward the larger online EI community. But in real life their response was more or less opposite. Brian retreated from the world until he could reestablish a reliable baseline of mental and physical health.

James's response led in the other direction. Concrete where Brian was metaphysical, he still believed that with sufficient exertion one could get to the bottom of it all, that there actually was a bottom. For him the problem was not perspective but *data*. If something couldn't be known, it was only because he lacked the right instrumentation. It was as if he were sunk in his own personal dark age, a time before microscopes, when all that was invisible to the naked eye remained unknowable, the subject of speculation, superstition, and fear.

James was luckier than most sensitives in that he still had money and could supplement his senses with technology. Thus the oxygen regulator, the heart rate monitor, and other gear. A year ago, after a mycotoxin contamination forced him to break a $7,500 lease on his apartment (picture him in downtown Santa Barbara, in a full hazmat suit, shuttling $5,000 worth of new clothes to a parking lot dumpster), he began looking around for even more sophisticated sensors. Eventually he located the Kestrel, but his hope was to use Arduino technology (an open-source platform for building custom digital devices) to one day develop a sensor that could also detect mold and chemicals.

Until then, when it came to housing James was just as screwed as every other sensitive. Residential real estate, after all, was premised on the assumption that a ten-minute tour was sufficient to determine the suitability of any given property, when for James the only sure way to know was to sleep there a week and note his reaction. Despite his resources, James, ultimately, could only refer back to his own feelings.

———

It's always revealing to see how people respond to uncertainty. By and large, the EI community had reacted by coming together. In fact, the more I learned about sensitives the more it seemed like they were, in their own way, flourishing. Not physically, to be sure, but collectively, culturally, as a community. They had their own medical treatments, their own referral networks, their own home building specifications, even their own dating service. Meanwhile the Internet positively bubbled with sensitive-friendly blogs and support groups where crowdsourced help abounded and sensitives could be assured of a sympathetic ear. Reading these sites you began to get a sense of the subculture that had quietly arisen around EI, in compensation for the lack of credence it received from the mainstream.

Indeed, the more seriously impacted sensitives had long since given up on fitting in. In place of this they had asserted a defiant pride in their nonconformity, embracing their sensitivities and dismissing non-sensitives as "normies." It was the same sort of pride that had buoyed the HIV/AIDS community—after they'd been ignored long enough. And enough people had died. True, EI couldn't kill you, but to be denied by society was like dying, in a way. Among its many other names, in fact, EI was also sometimes known as "Chemical AIDS."

But the same uncertainty that drew sensitives together could also lead to fractiousness and strife. I had already seen several such instances in the EI community—representatives of differing factions trying to discredit each other, for instance, or disgruntled sensitives sabotaging the living spaces of others: placing a flea bomb in a drop ceiling, or secretly installing plug-in air fresheners—the EI equivalent of an act of war.

The most acrimonious debates concerned the best way to treat EI. The options were ultimately as limited and as basic as our most primitive impulses: run, or fight. The runners ended up in remote locales like the desert Southwest, which, with its high elevation, dry air, and sparse population, had emerged as a promised land for sensitives. The fighters tried to stay in the game. As often as not this entailed acknowledging the psychological component of EI and confronting it head-on, usually with some form of "brain retraining."

Brain retraining was a commercial application of "neuroplasticity," a concept which had been around in some form or another since the 1960s, but really came into vogue in the 1990s. Basically, it was warmed-over cognitive behavioral therapy outfitted with some sort of neuro-lexical facelift. (The problem wasn't you; it was your amygdala. The goal wasn't to change your thinking; it was to "rewire your brain.")

But for many the snazzy language couldn't conceal the heretical imputation—that EI was at root psychological. And so the Facebook wars raged.

"Are there DVDs that will make the people in Flint Michigan not get sick from the water?" one sensitive from this faction wrote. "Seems along the same reasoning to me."

"Neurocognitive re-training is troubling for a lot of people," another added, more perceptively, "because it gives bad actors like defense attorneys and insurance companies and lazy doctors a huge out in how to victim-blame sick people for their condition and a way to completely dismiss the hard science."

A similar fractiousness could often be seen in the research community, where proponents of various theories flea-bombed each other in peer-reviewed journals. In one article, for instance, Pall ridiculed competing researchers by comparing their efforts to those of the fabled blind men trying to identify an elephant. Depending on where they stood, each had a different answer. Elsewhere, he responded to a conditioning theory of EI with even greater contempt.

"One is reminded," he wrote,

> in reading the Meulders et al. study, about the old English joke about a drunk at night on his hands and knees looking for his keys under a street lamp. An English bobby comes over and asks "What are you doing?" "Looking for my keys." "Are you sure you lost your keys here?" "No, I think I lost them over there," pointing to a very dark area a half block away. "Then, why are you looking here if you lost your keys over there?" "Because it's too dark over there."
>
> We can understand, in a sense, the perverse logic of the drunk. What we cannot understand is the insistence of Meulders et al. in looking for an understanding of MCS [Multiple Chemical Sensitivity] in dark speculation rather than in the bright light of science.

Pall's belligerence wasn't unusual. EI skeptics could be equally scornful and dismissive, relegating an entire body of research to the trash heap with a wave of the hand.

"EI advocates hide behind the cloak of uncertainty that enshrouds science," one writer proclaimed. "They profess to be scientists and researchers with a 'different opinion' and suggest an atmosphere of credibility. They present a facade of being dedicated researchers and clinicians excluded by the fraternity of the medical-industrial establishment. They view themselves as leaders of a populist movement. They do not discover scientific truths; they divine unfounded beliefs."

As with the biomarker arguments, the squabbling centered on the question of whether EI was "real" or "merely psychogenic." In the above writer's view, for instance, sensitives were motivated by nothing more than a fear of confronting reality and a desire to avoid responsibility, often stemming from a history of child abuse. But the tone of these exchanges suggested that there was a lot more at stake than opinions. The psychogenic faction left you feeling that they regarded the theories of the nonpsychogenic faction as an assault on science itself—and the solemn authority that went with it. As one outraged writer put it, EI was "the only ailment in existence in which the patient defines both the cause and the manifestations of his own condition."

The nonpsychogenic faction, meanwhile, felt that the EI skeptics, in placing a nitpicking empiricism before everything else, had lost sight of what the medical profession was supposed to be all about: the abatement of human suffering. In this sense the discord among researchers was merely a slightly more civil version of the fury and resentment that seethed among sensitives themselves. On blogs and online forums, I'd seen psychogenic researchers derided as "flat-earthers of the future," "the very definition of grasping-at-straws-failures," "shills for the chemical industry," and "frauds."

The level of animosity said less about EI than the anguish of being denied clinical legitimacy—whatever your ailment might be. In 2011, for instance, when a well-funded group of researchers published the results of a massive study suggesting that myalgic encephalomyelitis (ME) could be cured by a simple course of graded exercise and cognitive behavioral therapy, the ME community went ballistic. There were even reports that a coauthor of the study had been stalked by a woman who brought a knife

to one of his lectures. Another coauthor received a phone call from someone threatening to castrate him.

But even here I tended to side with the nonpsychogenic faction. Not
because of the science, which was never conclusive, but because, even if they
did err, they did so on the side of empathy, which at that moment in our
country seemed in egregiously short supply. And frankly the EI narrative
was already familiar to me, a medically coded version of the identity politics
that I had seen so many times before, from Frantz Fanon to transgender
rights, that I knew my part by heart and could sing along without thinking.

Thus the sarcasm of the skeptics, perhaps. It wasn't just a sign of their
impatience with what they regarded as bad science; it was a rebuke to the
ambitions of identity politics, a paternalistic attempt to limit what it could
claim. It was all very well and good to claim that you were male, female,
or other, for instance, but one thing you could not claim to be, no matter
how victimized you happened to feel, was sick. Not so long as *science* had
anything to say about it.

But there was more to it than that, for along with the sarcasm there
was also a long-suffering tone that could only signal exhaustion. It was as if
the skeptics experienced sensitives' medical theories as an imposition—not
just on medical resources but intellectual resources as well. The psychogenic explanation, after all, required no further research. Existing theory
was already sufficient to make sense of sensitives' experience. Accepting
EI as a legitimate disease, on the other hand, meant acknowledging ignorance regarding its cause. And acknowledging ignorance wasn't easy.

If the debate around EI revealed anything, it was just how much skill
and dedication being ignorant required. Even the people supposed to be
experts at ignorance—the scientists who spent their lives trying to eradi-
cate it—regularly succumbed to obfuscation, self-deception, territorialism,
and bias. After seeing enough of this one could be forgiven for wondering
whether our system of expertise was designed not to discover what wasn't
known but rather to uphold what already was. Confronted with unyield-
ing enigmas it turned temperamental—in medicine in particular. Nothing
made doctors grumpier than unexplained symptoms. Because medically
unexplained symptoms made them feel helpless. The easiest way clear

of that helplessness was to blame the patient. "Heartsink patients," they were called. "Malingerers." Which was more or less what the EI skeptics had concluded as well.

Perhaps it was foolish to think that the beef between the skeptics and believers could ever be neatly resolved. Both were laboring under an age-old misapprehension that human beings could be neatly divided between mind and body. Only rarely was the mind-body issue ever recognized in the scientific literature surrounding EI. It might not be that they were disagreeing, in other words, so much as speaking completely different languages. Pall himself claimed that even post-traumatic stress disorder, PTSD—another putatively psychogenic framework—could be explained by the NO/ONOO-cycle. And if PTSD, why not depression? Or anxiety? Or trauma from child abuse? Or any of the other psychological phenomena that skeptics cited to account for EI? Even supposing Pall was over-reaching, just because you could treat a disease psychologically did not (as Bell, to her credit, recognized back in 1992) mean the disease lacked a biological etiology. It might simply mean that the etiology was too complex, or too subtle, to detect. In which case calling something "psychogenic" was simply another way of declaring your ignorance.

The question, ultimately, was how, if, and where the biological and psychological frameworks converged. It was one of those questions that kept cropping up, no matter which way you turned. Certainly one could point to the expanding field of chronic pain research and ballooning scientific interest in myalgic encephalomyelitis as positive developments. The increased deference to the patient experience was another—as evidenced by the suddenly ubiquitous availability of alternative treatments, even in mainstream hospitals like Johns Hopkins, which offered Reiki to patients with digestive problems, or Duke, which offered detox programs and "botanical medicines" for children with autism and ADHD.

Even so, you couldn't help but feel that we weren't there yet, that some revolutionary new understanding of how the mind intersected with the body lay right around the corner. For surely in an era as technologically adept as ours the failure to understand the fundamental basis of our own nature could not endure for long.

CHAPTER 6

Down into the Country

Outside the Basalt Element, James finally finished loading the last of his princess pile of luggage and slid behind the wheel. Alternative rock crashed quietly in the background as the Rover came alive. James retrieved a boom arm from the backseat and affixed the suction cup end to the windshield. This was for the iPad. James plugged it in and quickly flicked through various maps and radar screens, demonstrating its utility for a runner like himself. Wearing a fresh pair of aviator sunglasses and a blue tennis pro visor, he looked breezy and capable, in fine health. He popped the lid on a two-ounce bottle of 5-hour Energy and took a swig. The biting, chemical reek of artificial pomegranate suffused the cabin.

"These are good to me," James said, holding up the little red and yellow bottle. "I used to love Red Bull but now it makes me sick. It has all this artificial stuff in it. 5-hour has mostly vitamins. Niacin. B_6. Folic Acid. B_{12}. Because I have vitamin B_{12} deficiency. And caffeine."

He eased the Rover back and we threaded our way toward the exit. At the stoplight, a homeless guy sat on a cement divider. His cardboard sign read "Homeless, Hopeful, Hungry, Humble," and it occurred to me that this guy could easily be another sensitive, blown out of his own life by the sudden onset of incomprehensible sickness.

It was a relief to get out of Basalt and go down into the country. The early light carved hard shadows on the hills, iron red and matted with sagebrush and gorse. As we drove they seemed to pace us like a pod of whales arcing against the pale horizon.

Down around Rifle the road straightened and the hills withdrew to a white distance, and the road shot forth under the blue sky. And within the Rover's cabin there tolled that long-awaited sense of arrival or perhaps *release* as we put our country and its politics and our own selves behind us and rolled into the vacant embrace of the land. James drove the speed limit, the road not asking much so he often piloted with his knees.

Around noon we crossed into the flatness of Utah, leached of green, shelterless and void. Strata of yellow and brown fell back from the black seam that ran through it, shadowed by distance and shot through with lavender and gray.

Somewhere outside Cisco we stopped to stretch our legs. Threading the empty land was a fragile chain of freight cars that looked like they'd been there for a thousand years.

"It feels a lot better to me out here than where we just were," James said, filling his lungs.

It would have felt better to anyone. Especially someone who was maybe operating with an impaired perspective. In the middle of nowhere, perspective was inescapable. In fact it's pretty much all there was. Every direction held the same yawning absence, the hostage taker's dream perimeter. Here nothing could creep up on you. Even the odd mountain lion or escaped prisoner could be spotted miles off. After all, it's not the enemy we fear most but not knowing the direction he'll come from.

Looking around, James remarked that he could probably buy up a lot of this land for pennies and use it to build a sensitive-friendly settlement, a Zion where you didn't have to worry about toxic building materials, tarry asphalt, or mercury in the drinking water. This sort of drive-by speculation was something James did often, an exercise in practical optimism, imagining perfect worlds.

James was already at work on his own perfect world, which could not have been more different than the desert around us. It was located on the north shore of Oahu, a ninety-five-acre oceanfront property he was developing into twenty-two condominium units. The Oahu project was not your typical slice-and-dice land development, designed to maximize profits at the expense of everything else. On the contrary, James was taking great pains to actively integrate conservation, agriculture, renewable energy, and historical preservation. Instead of leveling the old Marconi telegraph station erected on the property in 1914, for instance, he hoped to restore it and turn it into a high-end hotel. Instead of relying on fossil fuels, the land had been permitted for a 500kW solar array. Each two- to three-acre condominium parcel, furthermore, would include a farm plot for native fruit trees like pomelo, tamarillo, or lilikoi. In his conservation plan James even had the land blessed according to local Hawaiian custom.

"I may get some sheep, too," he said.

How he came by this property was a curious story. You could say it began in college, when he got caught up in the day trading craze and fell deep into debt. This was around 1996, at the height of the tech boom.

Desperate, James begged his father for a job subdividing a several-thousand-acre property that his father and his partner had recently acquired in Houston. His father agreed, and James moved to Houston, where he spent weeks living out of an unheated office building, subsisting on canned spaghetti. When the job was done James paid off his day trading debt and bought what remained of the property from his father to sell on his own.

Acknowledging that James had proved his mettle, his father let him go on a scouting trip to Oahu, where another large property had come up for sale. In the course of the trip James discovered several much more promising prospects. These had recently become available thanks to the demise of the local sugar industry, a colonial enterprise, initially, which had finally succumbed to high operating costs, overseas competition, and the artificial sweeteners which had been gaining market share ever since Monsanto introduced saccharin as its first product in 1901.

At this point James joined his father's company as a partner, and together they bought eight thousand acres in Hawaii. The partnership lasted eight years, until 2008, when James decided to go out on his own, taking a chunk of the Oahu property as his share in the company.

As someone still trying to make sense of James, several aspects of this story seemed noteworthy to me. The first was that James's father and his partner were serious businessmen. "Buying and selling is what they do," James said.

The second was that James's father didn't appear to give James any sort of preferential treatment. He refused to give James the Houston job at first, and only relented because his partner agreed to it. And even then he didn't make it easy. As James put it, describing those cold weeks in Houston, "I had to tough it out."

But I was most intrigued by the way James described his departure from his father's company in 2008.

"I was under a lot of stress," James said of his decision to leave, "and I thought it was the best thing to do."

When I pressed for clarification, his explanation deteriorated into tortured hesitations.

"I don't think there's anything wrong with the way they operate," he said finally, "but I think we just had different views about how to work on projects. It's just better to either be aligned, as one, or separate. That way, the decisions that I made would be mine, and the consequences, the results, would be either my fault or . . . it'd be my success or my failure. I'd be the only one to blame."

All of which struck me as a labored attempt to avoid saying something. It also left me wondering what James's father thought of his utopian Oahu project.

"The zoning may allow me to get a lot more units than I'm asking for," James had said, "but I don't wanna be one of those guys who comes from the mainland and cuts up the property and then leaves."

I suspected that James's father might be one of those guys. The Hawaiian term, James said, was *haole*—pronounced "howlie." "It means newcomer or white," he said. "It can be really derogatory depending on how it's said."

Was this what he was laboring not to say? That his father was a *haole*? The same way he refused to criticize Charlie when anyone with eyes could see he deserved it? Why was he so averse to placing blame when it was so well deserved? Was this a mark of his noble character? The same thing that kept him from ruthlessly exploiting the Oahu property? Or was it something else? A fear of inviting comment on his own shortcomings, perhaps, or evidence of his psychic domination by his father?

It was on Oahu, James said, that his EI was triggered by a toxic mold exposure. This was three years ago. He'd been staying in one of the Marconi buildings which he had fixed up to serve as a live-work space. But the tin roof was rusting out and he kept finding puddles on the concrete floor. He fixed the roof with a roll-on sealant, and while he was at it he replaced the windows and installed air conditioners, not realizing that sealing the building only trapped the moisture inside, where it festered in the Hawaiian heat.

Within a couple weeks he began to feel ill. At first, he dismissed it as allergies, which had always been a problem for him, or some combination of allergies and stress. But somehow this felt different.

"It sucked the life out of me," James said. "It felt like a struggle to breathe."

Thinking it might be the flu, he checked in with several doctors, one of whom diagnosed him with pneumonia.

The breaking point came four months later. James was in Waikiki with his mother, who was visiting for his fortieth birthday. He had just eaten lunch when suddenly he began to feel so bad that all he could do was stagger down to the beach and collapse on a towel.

"My neck was throbbing. My heart rate was way up and it felt like my artery was jumping. I felt like anybody who looked at me could see it. Just popping. And it was hard to breathe. I was thinking, 'This is crazy. I'm here in this beautiful place. And I may just have a heart attack right here, right on the beach, on my birthday!'"

Prompted by his mother, James called an ear, nose, and throat doc he knew in Beverly Hills, booked an appointment for the following morning, and got himself to the airport.

The ENT had no idea what was wrong with James. The mold theory he dismissed with a wave.

"That only happens to poor people in Pennsylvania," he said, and referred him to a pulmonologist. The wait was too long for the pulmonologist so James flew to Jacksonville, Florida, to see a pulmonologist at the Mayo Clinic. The Jacksonville pulmonologist ran a series of tests but found nothing. In fact, she told James that he was in the best shape of anyone she'd ever seen—physically, anyway—and advised him to try Flonase (an allergy med), Advair, and Albuterol (asthma meds).

None of which helped. James's family had a house in Florida, and he spent several days there just lying on the floor.

"My whole body hurt. My head, muscles, my joints. My leg was shaking out of control. And it was hard to breathe."

Finally, five months after the initial exposure, he decided to go to Denver. One of the best allergy centers in the country was located there. And he thought the clear mountain air might help.

It was there that James finally got online and began doing his own research.

———

Among sensitives, the most well-known research on EI was that of Claudia Miller, a professor of environmental and occupational medicine at the University of Texas Health Science Center in San Antonio. Miller's genius was to suggest that the controversy surrounding EI had less to do with the failure of EI to fit any known disease model than the failure of the existing disease models to accommodate EI. So she proposed a new one, which she called Toxicant-Induced Loss of Tolerance, or TILT.

Miller defined TILT as something like the inverse of addiction: it commenced with an exposure, and the exposure led to sensitivity—the opposite of tolerance. Thereafter, even minimal exposures could provoke outsized effects.

"Members of both groups," Miller wrote, "often report intense cravings and debilitating withdrawal symptoms. However, chemically sensitive patients' responses are not primarily to drugs. These individuals more commonly report addictions to caffeine or certain foods. While drug addicts manifest ad-dicted behaviors . . . chemically sensitive patients respond as though they were ab-dicted . . . and assiduously avoid the very substances addicted persons favor including alcohol, drugs and nicotine."

Miller didn't say much about how any of this actually happened, at one point labeling EI "a sort of 'cryptotoxicity.'" She did cite Bell's neural sensitization as one possible mechanism, but the primary contributions of TILT were nosological (pertaining to the taxonomy of disease) and conceptual. In this respect Miller was less a scientist than a marketer, packaging EI in a way that anyone could understand. Sensitives weren't crazy. They were just TILTed. Like a pinball machine.

"With a pinball machine," Miller wrote,

> a player has just so much latitude: he can jiggle the machine, nudge it, bump it, rock it, but when he exceeds the limit for that machine, the "TILT" message appears, the lights go out, and the ball cascades to the bottom. The machine's tolerance has been exceeded and no amount of effort will make the bumpers or flippers operate as they did before. The game is over.

TILT had much in common with "Rain Barrel" theory, a vernacular explanation for EI that often cropped up on sensitive-friendly websites. In this view, each individual had their own unique, genetically determined toxicity threshold. When that threshold was reached, the barrel overflowed, and the result was disease.

It was a pleasingly intuitive model. Pall's work might have been more rigorous, but it was this very rigor that made it all but impenetrable. In the process of explaining EI he had reduced it to a language so granular that no sensitive who wasn't also a cellular biologist could make sense of it. In this, Pall failed to do what (in medicine, anyway) mattered most: to console.

What made Miller a hero among sensitives, however, was not her sympathetic approach so much as her chutzpah. Far from being intimidated by skeptics who regarded EI as merely the ravings of attention-hungry malingerers, and EI apologists as their enablers, Miller argued that the real blunder belonged to the skeptics, for failing to notice that EI signified a whole new class of hitherto unexplained disorders, including ME, fibromyalgia, asthma, migraines, even Gulf War Illness, all of which her TILT theory could explain. In this, she upped the ante of the entire debate, in the process placing herself on the same footing as Robert Koch, the Nobel Prize–winning German microbiologist widely regarded as the father of germ theory.

"Just as the germ theory describes a class of diseases sharing the general mechanism of infection," Miller wrote,

> the TILT theory of disease posits a class of chemically induced disorders characterized by loss of tolerance to chemicals, food, drugs, and food and drug combinations. In the same way that fever is a symptom commonly associated with infectious diseases, chemical sensitivity may be a symptom associated with the TILT family of diseases. . . . The fact that this phenomenon does not fit already accepted mechanisms for disease is often offered as evidence that the condition does not exist. However, the same criticism would have applied to the germ and immune theories of disease when they first were proposed. What is possible depends on the biological knowledge of the time.

One had to admire her audacity. Especially as she went on to suggest that "the vituperative professional disputes that surround [EI]" were just one more sign that she was right. It called to mind the transcendent egoism of sensitives, appointing themselves prophets of an approaching toxic Ragnarök.

To support her claim, Miller invoked one of the biggest guns in the history of modern scientific thought, Thomas Kuhn. It was Kuhn who brought the term "paradigm shift" into popular usage with his 1962 book, *The Structure of Scientific Revolutions*. In it, he argued that changes in scientific thinking developed not linearly, in a steady, rational progression, but in a sequence of convulsions as old theories (Ptolemaic, Newtonian, humoral) succumbed to newer ones (Copernican, Relativistic, germ). One sign that such a convulsion might be near was the emergence of a growing number of anomalies that existing theories couldn't adequately explain. Like, for instance, EI.

In the context of EI, citing Kuhn was a devious way of suggesting that, since scientific theory advanced nonlinearly, it wasn't necessarily reasonable to expect that EI should conform to existing theory. In fact, the seeming irrationality of EI might even be a sign of its legitimacy.

There was a kind of heedless anarchism and effrontery to this line of thinking. You could almost picture the EI skeptics pulling out what remained of their hair. In a way, the entire setup felt like an experiment designed to assess one's underlying sympathies. The credibility question again, but reversed to implicate the onlooker. Where did you stand? With the respectable, conservative advocates of reason and the scientific method? Or the disreputable outcasts who felt like something about the way we were living was seriously, deeply wrong?

Not coincidentally, TILT itself had a lot in common with Kuhn's model. Both were nonlinear, premised on thresholds. Both were contrarian and disruptive. With TILT, seemingly irrational symptoms were proposed to be rational. With Kuhn, seemingly rational science was proposed to be irrational. If nothing else the tonal similarities explained why both proposals were met with a certain degree of outrage.

Perhaps the most inflammatory aspect of Miller's reference to Kuhn was the suggestion that scientific knowledge was not uniformly accessible and public, as so many liked to think, but private and highly structured, the proprietary domain of those with the greatest stake—hospitals, universities, research centers, insurance companies, chemical companies, and government agencies, among others. Indeed, knowledge was if anything a kind of property, and like any other property was subject to the biases of whoever owned it. In this sense the controversy over EI could be seen as something like a property battle between the current "landowners" and the insurgent rabble who aimed to claim the property for themselves.

It's this contested framework that made the ownership of knowledge subject to convulsive change. And from here it was no great leap to notice that the likelihood of knowledge changing hands was directly related to how much the current owner was invested in it. This explained much of Miller's rhetoric, inveighing against "carpet and rug manufacturers, fragrance manufacturers, pesticide producers, building owners' associations" that hired "physicians and researchers as expert witnesses," as well as pharmaceutical companies that were "often owned by chemical corporations," even the National Institutes of Health. Not to mention the "37,000 psychiatrists and 241,000 psychologists in the United States" who all but guaranteed that any theory suggesting "that depression, anxiety, panic attacks or fatigue might be caused by chemical exposures" would get "a less than enthusiastic reception."

In framing the battle over EI in such broad terms, alongside the by now familiar and perfectly credible data about "the exponential increases in synthetic organic chemical and pesticide production that have taken place in this country since World War II," Miller offered an entire worldview in which the paradigm shift had, in effect, already happened. In fact, she suggested, everything was shifting, from the quality of our environment to the explosive growth of the chemical companies responsible for its degradation; from the power and influence of pharmaceutical and insurance industries to the ability of government agencies to regulate them. Yet despite all these shifts our medical sense of ourselves still somehow lagged behind. In this sense what Miller was really offering was an account of how we failed to perceive ourselves, failed to know ourselves. It was a failure not of knowledge but of self-knowledge. And it was from this failure, so the thinking went, that all other failures were allowed to proceed.

Naturally the solution had to be as dramatic as the diagnosis. Desperate times called for desperate measures—a revolution in medical thinking of a sort not seen since the advent of germ theory. The claim was extravagant, but Miller disguised it by putting it in the service of a greater humility. In the Civil War, she wrote, hundreds of thousands of soldiers died from infections that, at the time, were attributed to "miasma." "It is possible," she wrote, "that we may be at the Civil War stage in our understanding of chemical sensitivity."

There was a certain perverse appeal in this. The idea that we might be far more ignorant than we presumed. It felt right—especially to anyone for whom the modern threat landscape had become more or less unreadable. It returned you to the idea that sometimes the greatest obstacle to understanding was not necessarily ignorance but the reluctance to acknowledge it.

More broadly, it echoed the environmentalist motif of human hubris and its ecological consequences, and the attendant view that what we needed was a way to recognize our place within the larger scheme of things. What we needed was to be humbled, like that dude on the cement divider. Like Job.

It was a good story, with an intuitive appeal. And as any marketer knows, nothing shifts paradigms like a good story. That is, so long as no data is available.

Interestingly, Miller was not entirely alone in her view that EI might represent a new model of disease. Pall also compared his work to Koch's at one point. And for all anyone knew, either or both of them might have been right.

CHAPTER 7

The Word from Snowflake

That first day out of Basalt we drove for eleven hours. It was around two when, somewhere in Utah, we reached the edge of one of those rare zones entirely free of human commerce. NO SERVICES NEXT 206 MILES, a sign warned.

The land was changing. It was as if we were emerging from a dry seabed, the alkaline plains sifted by wind and drifting with bone dust rising up so the road cut right through in places, and you could see the earth's chalky interior, the hard rime of millennia which left you feeling vaguely guilty or disrespectful or maybe just frivolous, like a child among adults whose deep timbres you cannot decode. Emerging from these cleavages the eye would shift to other magnitudes as the land fell away to a calm haze, as if the terrain in those far places had yet to be fully assembled. Soon great piles of rock began to appear, wrinkled battlements surrounded by the dross of their own crumble the way an autumn tree is ringed by leaves.

We pulled over for lunch at a dusty turnout in the lee of a towering pile of red rubble, the wreckage of something never built. Outside the car you could feel the dry air prickling your pores. The heat was a steady bake that made you want to hold lizard still as if that way the sun would bc lcss likely to notice you, this body with its rushing red rivers that it presumed to keep all for itself.

James clicked open the tailgate and raised the rear hatch to make some shade and we unpacked our lunch from Whole Foods. We had both left messages for Brian and were still hoping to hear from him, for while James knew the forest where Brian was staying he did not know the exact mile marker.

James tried his number again but got no answer, and I flashed on an image of Brian's remains, propped on a rock somewhere, clothes falling off and shins gnawed by coyotes.

Of course, it was possible that Brian's phone had simply run out of juice, or been scuttled by a flood. In part it was a question of how seriously to take the concern expressed in the Facebook groups. It seemed like they were always overstating one peril or other. But maybe I was only telling myself this to justify the detour to Snowflake.

Both destinations exerted their own draw. Brian was the most far-flung. If the Snowflakers were pariahs, Brian was one step beyond. He was like the last man, the closest to the void. I had framed it as a humanitarian mission, but really I wanted to hear from him what the void had to say. And, I suppose, in a more macabre sense, to discover what distillate remained after the Jobean trauma had reduced everything to ash.

With Snowflake the appeal was different. Like everyone else I was accustomed to the constant renegade posturing our culture indulged in—the articles titled "I Went Offline for a Month and Here's What Happened," the cinematic fantasies of blowing up credit card companies or living off the land—but I was fascinated to see what a community of true apostates looked like. Folks who truly had cut loose from the mother ship and renounced the regime of pleasure and convenience that held the rest of us captive.

I suspect my interest was not that different from that of the typical curiosity seeker—with the exception that I was ready to own my curiosity, and all the ways that it compromised me. Or thought I was, anyway.

Snowflake began with a man named Bruce McCreary. It was McCreary who gave the most cogent explanation of what EI felt like that I had ever heard.

"It feels to me like there's a little three-year-old going *Hey. Hey. Hey*," he said, "only it's in my head. I mean not talking but constantly interrupting . . . You know how you feel when you're constantly interrupted and you can't concentrate, because you keep getting interrupted? Only, I self-interrupt."

McCreary was an electrical engineer by training. He became chemically sensitive after a workplace accident involving solvents. Twenty-nine at the time, he took to the road in search of an environment he could tolerate, and eventually settled in Snowflake. At nearly six thousand feet above sea level, with dry air, sparse vegetation, and little pollution, Snowflake had a lot to recommend it.

Over time, McCreary reached out to other sensitives and convinced them to join him. One of these was Susan Molloy, who became sensitive at age thirty-one, and attributed the onset to childhood pesticide exposures in Oregon, where she grew up harvesting beans and strawberries to pay for her school clothes. A disability activist, Molloy became a leader in the Snowflake community, letting newcomers camp in her driveway, handling correspondence for those who couldn't bear to be near computers, and sometimes shopping for those who couldn't tolerate stores. Molloy had done much to establish Snowflake's nonjudgmental culture, and she later became something like the community's spokesperson.

McCreary, meanwhile, having been trained as an engineer, focused on developing practical solutions to all the unusual challenges EI presented. This was part of what interested James, for he, too, imagined building a sensitive-friendly community one day, once his Hawaiian Eden had been achieved, and he was eager to see how the place worked.

McCreary had taken particular pains to outfit his home to accommodate his sensitivity to electromagnetic fields (EMFs), adapting his refrigerator to emit zero radiation and his power tools to run pneumatically. In time, he developed a complete EI home building protocol—most of which entailed construction techniques that avoided toxic building materials.

The protocol detailed everything from the wood to be used (kiln-dried hemlock and Douglas fir, if maple and poplar was too expensive) and the sort of kitchen cabinetry (steel with a baked-on epoxy coating) to a technique for sealing the walls with industrial-grade aluminum foil glued on like wallpaper with a wheat paste adhesive and painted with an experimental clay paint McCreary had formulated.

Many in the EI community regarded Molloy and McCreary as empowering figures. Their efforts signaled a healthy self-reliance, and a clear-eyed acceptance that sensitives could count on no help or sympathy from the mainstream.

But others (in an echo of the brain retraining debate) wondered whether, in adapting their homes—and their social mores—to better suit their sensitivities, the Snowflakers were not creating a home for themselves so much as for the disease.

And there was reason to think that this might be true. Suicides happened as often as twice a year in Snowflake, which was significant, given that the community consisted of only about thirty households. People had been known to starve themselves.

The one woman I talked to who managed to leave Snowflake after living there for five years spoke of her departure almost like an escape.

"I left the protective life that I was living," she said. "I sold everything. My oxygen tanks, my masks, whatever I had that I thought was protecting me. And I just took a risk . . . Because living the way I was living was killing me."

It was this faction that had been trying to discourage me from visiting Snowflake, on the grounds that the people living there weren't representative of EI in general.

"Including Snowflake in your story would be similar to writing an article about marriage and including a passage on Warren Jeffs," one woman wrote, "that leader of a polygamist cult who was taking lots of very young girls as brides. Yes, it's the end of the spectrum of the story of marriage, but not one that needs to be included."

At any rate, I couldn't get through to anyone at Snowflake. Susan Molloy hadn't responded to emails or phone calls. I'd reached out to various other residents but so far had received no response. So I was surprised to hear my phone break the hot silence as we sat on the tailgate and find a woman from Snowflake on the line, her voice cold and distant.

━━━

On the road again we tacked south toward Goblin Valley, flanking the great Capitol Reef, a 200-million-year-old geologic overbite running from Thousand Lake Mountain in south-central Utah all the way down to Lake Powell, near the Arizona border. A faint fizz of life clung low to the ground here, the merest whisper of it like the sound of a crab on a beach. It felt like the surface of an alien planet, the kind of land you might see a rover picking over at two miles an hour, pausing to fondle a rock with a titanium paw.

To the east the land began to rise and acquire features, sandstone bluffs a man might aim his horse toward in hopes of finding water. Eventually there took shape on the horizon something that looked like a giant stone rabbit, red in hue, nibbling the thin vegetation. The road dropped toward it like a runway, the land so empty it made everything seem to stand still, no matter how fast you were going. James kept it to a steady 75, nudging the wheel with his knees.

The call had been from someone named Liz. It was clear she had been elected to vet me. We spoke for fifteen minutes and in the end agreed that she would consult with several other Snowflakers and maybe call back. There was nothing to do but wait and hope that they would give us a chance, though I wouldn't have blamed them if they didn't. To them, even James—still at large in the world, and fighting for his right to be in it—was a different breed.

Among the things that stood out in the conversation with Liz was her insistence that her condition be regarded as a kind of traumatic brain injury, or TBI, no different than the sort of injury football players routinely suffered. I mentioned this to James, and he picked up on it with his usual boyish enthusiasm, affirming the term on the grounds that hyperbaric chambers, one of the few treatments that afforded him any relief, were also often used to treat TBI.

I continued to be startled by James's readiness to entertain offbeat ideas. You could see examples of this readiness in the wide range of treatments he had undergone—like the hydrogen peroxide, which he administered with a nebulizer as well as by the drop. The nebulizer he originally acquired to deliver glutathione, a supplement recommended by one of his doctors, Dr. Hope. James had seen such success with hydrogen peroxide taken internally that when he developed a chest cold one day he decided to try inhaling it. It burned a bit, but seemed to work. Later, when he developed a case of pink eye, he decided to try it again.

"I just held the nebulizer up and rolled my eye around," he said. "And it went away!"

He said this with a kind of magical flourish—poof!—that reminded me of . . . what? A kid with a science kit? In James's world, you never knew what you might discover.

He had used hydrogen peroxide to treat his cold sores as well. Terribly painful, he said. But it worked. For a while. He had better luck with prolozone, which was basically injectable ozone. He had learned about it while investigating possible cures for rheumatoid arthritis, which affected his lower back and right leg, and tended to act up when his EI was triggered.

A naturopath in Prescott, Arizona, gave him eight prolozone injections. James was so pleased with the results that he asked the doctor to inject the prolozone into his lip to treat his cold sores.

"The next day they were gone!"

Poof.

I first thought that James's open-mindedness was simply an extension of the same studied nonjudgmentalism that kept him from condemning Charlie—or his father, for that matter, if the shoe fit. But the way James told it, he hadn't always been so receptive. The key moment came in Sedona in late 2014, around the same time as the mountain lion encounter. By this point he'd been in and out of hotels and Vrbos—vacation rentals by owner—for over a year, and seen countless doctors. Among them only Dr. Hope had acknowledged that his symptoms were real.

"That was the point where I think I was at my lowest," James said. "I was just thinking, I'm gonna go to Sedona, I'm gonna rent this house, and

I'm gonna do everything Dr. Hope says. And I'm gonna try some things that are . . . different. Whatever it takes. I'm gonna get better or die."

Thereafter, James would take his cues primarily from how his body responded to various treatments. Every success signaled another set of dots connected in a world where connections were few and far between. As time went on, his appetite for these connections led him to try anything even remotely credible, from autohemotherapy and chelation to magnesium baths and salt caves. It was unknown terrain, but at least it offered room for hope.

It didn't come without a cost. Like suddenly it became a whole lot easier for people to dismiss him as crazy—including his own family. Nor was it easier sticking to a single fixed address in Sedona. On the road it always felt like things were changing, like the answer might lie just around the next bend. Even loneliness felt more bearable on the road. In Sedona by contrast James woke up to the same loneliness day after day: a large, empty house borrowed from a stranger in a town where he knew no one. Sedona was a friendly city, but even so it felt like exile.

The loneliness eventually led him to Facebook, where he discovered a sprawling fellowship of other sensitives all equally exiled but fervently communing online. There were a number of different groups, some with members in the thousands, and their tone was first and foremost supportive. In a sense, it was the best of the Internet, a social safety net for people who had been spurned by their doctors and their families, a way for isolated sufferers to hear what no one else was telling them: *You are not a freak.*

The groups rang with cris de coeur and heartfelt empathy. Deep friendships were made here—including James's friendship with Brian—even though the vast majority of people had never actually met.

There was, as well, no end of practical advice, shared with the avidity of hobbyists. Tips on the best brand of shower filter, natural fiber mattress, or heavy-metal-free tattoo ink. Tips on how to heal the kidneys using only baking soda, how to patch a leaky blood-brain barrier, or how to pop a sealed hotel window with a screwdriver. Tips on which supplements offset thyroid deficiency, which department stores had the least VOCs—volatile organic compounds—which coastal towns in central California had the best airflow. Not to mention reliable washer-dryer brands, like the Speed Queen that James purchased on Brian's advice, after the previous seven didn't work out.

"Have you tried zeolite?" one person might inquire, in response to a question regarding how to positively acetylate a cofactor known as NADPH.

"Yes," would come the response, "I have found that taking a scoop of zeolite and a scoop of microsilica is amazingly helpful for me. Oxalate dumping mobilized a lot of mercury for me."

Reading these threads, which sometimes topped a thousand comments, you began to get a sense of how much energy sensitives spent trying to achieve the homeostasis that most of us took for granted. It left you feeling that the greatest gift of all is not having to pay much attention to yourself. This is what illness did, is make you notice yourself in unwelcome ways.

In the Facebook groups, there was no end to this noticing, no end to the lengths people would go to try to understand. The impulse could only be checked by achieving some kind of certainty, and the lack of instrumentation made certainty impossible. So the search became ever more obsessive and strange.

Discussions about locations in particular exhibited an almost animal sensitivity. In this, Brian's contributions were exemplary.

"I found Scottsdale to be feast or famine," he had written at one point. "It's a bit more elevated above the Phoenix settled toxins, if memory serves . . . I found some good micro climates between Cave Creek and that main drag, where it's a bit more elevated too . . . Every low pressure system makes every area down there insufferable."

"Cave Creek has cancer clusters," someone else chimed in. "Tons of brain cancer and especially in children. There's also a nuclear plant by Glendale."

"St. George, UT, where I'm at, has massive cancer clusters from radioactive nuclear waste in the ground here," came the response. "Maybe that's what I smell and feel here."

Undergirding all of this was an understanding that, not only was certainty impossible to achieve, but solutions varied from person to person. Some people thrived in coastal climates. Others thrived on the desert plateau. Some people swore by calendula oil. Others had good luck with Mosapride, VitaBlue Minerals, or frankincense. Some people could tolerate rice wine. For others, it was Blaum Bros. moonshine, or Angry Orchard cider. There was no single answer, just as there was no single body, no Vitruvian man. The only thing everyone seemed to agree on was that everyone was entitled to be sick in their own way.

Here was the identity politics that drove skeptics so crazy. For them, the diagnosis and treatment of illness was not a matter of personal preference but objectively verifiable science. To presume otherwise was to invite distortions arising from the observer effect—a feedback loop wherein the act of observation alters that which is observed, until it becomes impossible to distinguish what is actually there from what might have been conjured by the observer.

In the Facebook groups, the observer effect often led to speculative free-for-alls. In one case, for instance, a woman became convinced that a supplement suggested to help regrow her broken toenail was actually helping her mood.

"I thought this very strange," she admitted, "(why would silica make me happy?) and figured it must be the placebo effect, or maybe just feeling empowered by actually DOING something (my toe has me just lying here, not feeling great). But then I did a little research . . . I think it's the choline! Maybe I needed some, after all?"

CHAPTER 8

The Work of the Devil

We still hadn't heard back from Brian or Liz but reaching either meant crossing the same desolation. It looked like UFO country. The kind of unwitnessed place where the improbable would choose to occur. As if in verification of this we kept seeing trucks go by towing boats. Bound for where? The last great body of water to touch this dust was 250 million years ago. Once a jungle thrived here. Now there was nothing.

At around 3 p.m. we hurtled over an arid ditch called the Dirty Devil River. The speed limit had risen to 80 but it might as well have been 200, because, somehow, the red rabbit was getting no closer.

Now, after six hours on the road, a weightless silence descended between us. It was the kind of silence that marked a shift in a relationship. James might still be a puzzle in some respects but I was pretty sure he wasn't a lunatic. One thing did still trouble me, however.

The Glock. I had researched the gun back in Basalt. Among enthusiasts, this model—"the most powerful Glock pistol in existence"—was known as "Glockzilla."

I do not like guns. I may not be able to swing belief in God or any other higher power but surely guns violate some universal law of scale or proportion, and to my mind this made them as close as you could get to unholy. Perhaps I was just a mild-mannered New Englander, but to me even holding a gun felt like gripping someone by the neck.

In this, the Glock had opposite meanings for James and me. For him, it was a protection against threats. For me, it *was* the threat—if only because, not being able to see myself on one side of the gun, where else could I end up but the other?

Perhaps I wouldn't have felt so spooked if the gun had appeared at the end of the road trip. When a gun shows up in the beginning you can't help worrying (per the old thespian rule) that it will return later in less friendly circumstances.

Which is why, deep in the no-service zone, I suggested we pull over for some target practice. Better, I thought, to bring the Glock out now and take the spook off it.

James met this suggestion with his usual exuberance. We stopped by what looked like a heap of waste tailings but could just as easily have been a natural feature of the landscape. The ground was a giraffe hide of parched polygons, the edges curled by the heat.

Popping the Rover doors we stepped out of the cabin like astronauts and were met by stunned silence and dust. We looked around for something to shoot but the land was impervious, yielding nothing.

Finally James pulled an empty Fiji bottle from the backseat and threw it skittering across the hardpan. He drew the Glock from its Kydex holster and aimed.

Crack. Crack crack crack. Crack crack crack crack crack crack.

"Wanna try?"

I took the gun and ignorantly aimed it and pulled the trigger. It was a strange feeling. Like manually operating a metaphor. Like holding a car crash in your hands. The Fiji bottle leapt into the air like a startled cat. What, I wondered, would it feel like to shoot a mountain lion?

When the clip was empty James retrieved the bottle and like infants we examined the punctures we had made. It was the first time James had used the gun and he seemed well pleased with it. More powerful than a 9mm, he said, but still accurate.

"One thing I didn't notice until I went to shoot was the tremors," he said, holding out his hand. "A month and a half ago, without any myco-toxin exposure, I'd have no tremors. But as soon as I put the sight on the bottle, I could see it then. It was, like, shaking."

I didn't doubt the tremors, but it was hard to know if they were due to the mycotoxins or the 5-hours James had been pounding all day. This was the thing about EI. It made you see symptoms everywhere. Just like the gun made you see targets.

——

It was exactly this sort of perceptual warp that prompted EI skeptics to cite the higher rates in mental illness that accompanied EI, particularly delusional disorders, OCD, and anxiety. The question was which came first. Did the mental illness give rise to the EI or was it the other way round? Living with a disease as elusive as EI would make anyone at least a little loopy, after all. In fact, elevated rates of depression and anxiety had been widely noted among those who lived with other types of chronic illness.

Besides, what seemed like unhinged reasoning to others might, from a sensitive's point of view, simply be a different mode of thought. When the old way of thinking about things failed to yield results, the sensitive (like any pragmatic, solution-oriented person) tried something different. What a sensitive's baffled family member mistook for insanity, in other words, might actually be the sensitive's attempt to hang on to what sanity remained.

Thus James abandoned his skepticism, acquiescing to a world that for him had already acquired certain quasi-magical properties and adopting a superstitious style of thinking to suit. In this light, superstition was simply what happened when one's need for knowledge outstripped one's ability to get it.

The relationship between uncertainty and superstition was well docu-mented by research: the more anxious you felt, or out of control, the more likely you were to identify patterns that weren't there. In one experiment, researchers induced anxiety by furnishing subjects with random performance feedback. From the study it wasn't clear what this was, but you could imagine participants being asked to perform a task, like drawing a picture or whistling a tune, and then being solemnly assigned completely random scores for their performance.

Having thus messed with their subjects' heads, researchers then exposed them to snowy images, like TV static, and asked them to describe what they saw. Subjects in whom feelings of anxiety or a lack of control had been induced were more likely to perceive images where none existed.

It was only rational. When something didn't make sense you tried harder to make sense of it—particularly if a lot depended on your doing so. By the same token, the same study noted, the deep-sea fishing tribes of the Trobriand Islands, where sudden storms were a constant peril, had far more fishing rituals than tribes who kept to shallower waters; parachute jumpers were more likely to see a nonexistent figure in a picture of visual noise pre-jump than post-; baseball players who were more exposed to unpredictable play (pitchers) had more baseball rituals than players who were farther away (outfielders); and first-year MBA students were more susceptible to conspiratorial perceptions than second-year students.

Perhaps the most common example of this sort of behavior among sensitives was a theory cooked up to explain the most virulent and tenacious toxin ever. "Hell toxin," they called it.

"We figured it was HT," one sensitive wrote, "because it spreads faster than anything I have ever seen, explodes like crazy on contact with water, hates chemicals, and makes us very sick. It's the most pernicious toxin we have ever encountered. If it's not HT, it's probably HT's cousin."

The only way to get rid of hell toxin was with a blowtorch. I'd heard one story of a woman who had actually done this, literally blowtorched the walls of her apartment.

The term "hell toxin" had been introduced in early 2014 as EI continued to bulk up on terminology that was being withheld by mainstream medicine. (Another example was something called "Desert Gaggy": "the dust and moisture odor of many houses in the desert that have had swamp coolers"—or, more broadly, the smell of a place you don't want to live). Two years later the term had gained wide acceptance among sensitives—moldies in particular—and even had its own fiercely private study group on Facebook.

Yet, unlike EI itself—which, though controversial, was accepted as an actual syndrome by a minority of ecologically inclined researchers—there was no academic recognition of or clinical evidence for hell toxin. The desperate need sensitives felt for control in their lives had simply compelled them to make it up.

"Having had repeated encounters with this one," the study group founder wrote, "my strong view is that if you have encountered it, you know it. You don't wonder."

The hyperbole of the term itself was revealing. Extreme experiences required extreme terminology. It was perhaps not surprising that the terminology sensitives came up with obliquely referenced religion. As far back as the thirteenth century diseases that could not be identified were assumed to be the work of the devil. All that had changed since then was the frontier of what was known. The devil remained exactly where he had always been, directly across that frontier, ready to take credit for everything else.

Brian himself sometimes used the term "Biblical Toxin" ("Not to be confused with the lesser evil 'Hell Toxin'"). He once even underwent an actual exorcism. ("I'll never forget that 80 degree February day on that balcony, getting screamed over by an over-cologned, unabashedly egotistical 'faith healer.' I was thinking, 'So this is what it's come to?'")

Which explains why at least one skeptical toxicologist went beyond the mental health slurs to compare EI to actual witchcraft. Like witchcraft, he wrote, EI provided "an explanatory frame of reference which removes uncertainties and prescribes steps for the management of the tension." What's more, he argued, whenever contradictions arose, they could always be explained away with what Michael Polanyi (another heavyweight from the history of modern scientific thought) called "epicyclical elaboration." The term referred to the elaborate mathematical excuses Ptolemy offered to account for the failure of his geocentric astronomical model to explain the behavior of the planets. Thus, when a certain toxin failed to trigger the same reaction in one sensitive as another—or even in the same sensitive at different times—the inconsistency could always be attributed to variant airflow, genetics, or a number of other convenient variables.

The skeptical toxicologist had a point, but this alone did not give him the right to dismiss the idea of hell toxin—and sensitives themselves—out of hand. Hell toxin was ultimately just a colorful placeholder for something sensitives didn't fully understand. Such placeholders were common in the scientific community. I could think of several instances, for instance, in which scientists intuited the existence of an element or subatomic particle that they hadn't yet actually discovered. Giving that unfilled space a name helped scientists orient their thinking. In the face of the unknown, it (like witchcraft itself) enabled sane behavior.

Polanyi was kind of an unfortunate choice, heavyweight-wise, because among other things he had also argued that modern science itself was often guilty of the same sort of epicyclical elaboration as Ptolemy. "Secured by its circularity and defended further by its epicyclical reserves," Polanyi wrote, "science may deny, or at least cast aside as of no scientific interest, whole ranges of experience which to the unscientific mind appear both massive and vital." Polanyi then went on to invoke a heavyweight of his own—William James: "We feel neither curiosity nor wonder," James wrote, "concerning things so far beyond us that we have no concepts to refer them to or standards by which to measure them." By way of example Polanyi cited Darwin's story of the Fuegians, an indigenous tribe on Tierra del Fuego, who stared in amazement at the little rowboats that brought the sailors to shore, but paid no attention whatsoever to the great sailing ship from which they were launched.

What's more, Polanyi argued, scientists always make leaps (he had a great influence on Thomas Kuhn). According to Polanyi, for example, Copernicus's assertion that the earth orbited the sun was not the fruit of some laborious calculation but simply due to the fact that Copernicus found the idea intellectually pleasing. "In a literal sense, therefore," Polanyi wrote, scandalously, "the new Copernican system was as anthropocentric as the

Ptolemaic view, the difference being merely that it preferred to satisfy a different human affection."

All of which sort of left you wondering about the value of scientific accuracy. What was more important? That you have some grasp of the world? Or that your grasp be scientifically correct? A researcher would probably go with the latter, but a patient, a patient experiencing real suffering, might well disagree. Only someone who was used to having some grasp of the world—the epistemologically spoiled, you might say—would underestimate how important that grasp was. And how difficult it was to live without. As one sensitive said, "If I'm a kook, OK, at least I feel better."

That said, there was definitely a point where what you might call permissible anthropocentrism turned into something else. With so little known about what ailed them, sensitives worked with what they had—building links, noting similarities, connections. If the causes of two phenomena were equally unknown, for instance, it followed that they must be related. Thus, when a news report emerged about bacteria in Kazakhstan triggered by strange weather, killing 200,000 antelope in the span of three weeks, the moderator of one group wondered, "Do you think this is part of our phenomenon?" It was the same sort of thinking behind Pall's belief that his NO/ONOO-cycle could explain not just EI but also pretty much every other mysterious ailment out there. It was the same sort of thinking that gave rise to conspiracy theories: chemical companies in league with the government; the medical establishment in league with insurance companies. As for the cell phone industry, they regularly spiked legitimate safety studies to hide the truth.

"It's like a Jason Bourne movie," one sensitive said. "The number of people who come up with these studies, or are fighting the cell phone industry . . . there's a spectacularly high number of fatal accidents, sudden heart attacks, houses burning down."

One theory held that someone was killing holistic medicine doctors. Why not? The medical establishment clearly regarded holistic medicine with cynical ill will. And if holistic medicine ever gained hegemonic supremacy, its prevention-based approach gradually edging out more expensive interventional practices, the backers of mainstream medicine stood to lose billions.

This particular narrative began with the death of Dr. Jeff Bradstreet, a highly visible anti-vaxxer (he had twice testified before the House of Representatives) who had been raided by the FBI in 2015, and shortly thereafter was found dead on the bank of a North Carolina river with a gunshot wound to the chest. This was more than enough to get the ball rolling. Subsequently, each time a holistic doctor died under even remotely suspicious circumstances their name was added to the list. By the time I went searching for Brian the list numbered around sixty.

"There is no such thing as coincidence," one sensitive observed.

"MK Ultra style plus antidepressants," wrote another. "Common link often is—mentally disturbed person is screwed up on pills and becomes a hypnotized puppet. Who are the controller hypnotists and how long term is their game plan?"

In the Facebook groups questions of this sort appeared right alongside questions about nontoxic contact lens cleaning solutions, or vacuums with an exhaust attachment that could run outside. The unwritten rules were just as accommodating as mainstream medicine was unaccommodating. Collectively, it amounted to a kind of anarchic revolt against the expertise that had left sensitives high and dry.

We were already living in an era when the very idea of expertise had never been more distrusted. Books routinely appeared with a lot of yap about the hubris of coastal elites, the overvaluation of a college education, Internet echo chambers, the marginalization of mainstream news sources, and the rise of crowdsourcing, climate change deniers, and self-help. Trump himself was, if nothing else, a profound expression of the loss of faith in political expertise.

It's worth noting that this revolt against expertise was not Kuhnian in nature, not a case of one paradigm being overthrown by another. It was simply that experts across multiple fields had become so compromised by the very structures that had been put in place to bear their expertise that they could no longer be trusted to do their jobs. Politicians didn't serve their constituents; they served the machine. Doctors didn't serve their patients; they served the insurance industry. Scientists didn't serve the truth; they served the viciously competitive world of scientific research. And something similar could be said of the news media. Even our religious leaders, in covering for chronic sex offenders, repeatedly placed the reputation of the Church before the interests of those whom the Church was ostensibly created to serve.

Arguably, this was less a problem of human weakness (human nature, after all, would always be the constant in this equation) than of scale. Once the sustaining structure of expertise got too big, it warped the purpose it was built to serve. The obvious solution was to limit the scale, shrink the structure, and bring the humans that had been occluded by that structure back into view.

But how likely was this? Sensitives were not optimistic. In fact, they were even less optimistic about medical expertise than most people were about the news media, or government. Their skepticism was heightened by the sexism that off-gassed from the medical establishment. Some doctors, for instance, interpreted the gender imbalance in EI as evidence that it could be attributed to the sensitivity of the female character, despite the nonuniqueness of this gender imbalance to EI, and the dozen other possible things that might explain it—e.g., the interaction of chemicals with estrogen; women's smaller body size; their greater percentage of body fat in which chemicals could accumulate; their lower body concentrations of alcohol dehydrogenase (an enzyme which detoxifies carbohydrates, sugar, alcohol, and chemicals); their greater use of chemical fragrances, cleaning products, and cosmetics; their greater likeliness to hold low-paying jobs with poor air quality; the fact that women are less susceptible to the macho impulse to conceal symptoms; and so on.

I'd heard plenty of stories of doctors condescending to their female patients, writing off their symptoms as psychological in nature and leaving them feeling ignored, belittled, or blamed. One doctor suggested "wine and chocolates" as an appropriate cure. There was nothing new in this. Medicine had a rich history of patriarchal bias. In the nineteenth century, when it was more prevalent among women, even cancer was once attributed to their "greater feebleness." Since then, studies had repeatedly shown that women were less likely to be taken seriously by doctors. The bias was built in. Pharmaceutical dosage guidelines, for instance, were rarely broken down by gender, despite obvious differences in male and female biology. And despite advances in chronic pain research, and the fact that 70 percent of chronic pain patients were women, 80 percent of pain studies were conducted on the cells of male subjects. It wasn't until 2016 that the NIH began requiring its research grantees to study both sexes.

Indeed, "expertise" and "patriarchy" had become almost synonymous. And so, as women increasingly called patriarchal attitudes into question, and terms like "mansplaining" worked their way through the culture, leaving a trail of upset presumptions in their wake, the possibility of expertise was called into question as well. There was the sense that it might need to be completely reformulated. For sensitives, this reformulation began much as it did with abortion activists forty-five years prior: with their bodies, which they would always know in a way that no expert could—regardless of how powerful their scanners might be.

This explained the curious permissiveness of the Facebook groups. It's not that everyone had simply decided to be friendly and supportive; it's that the usual basis for making claims about the bodies of others had been rejected as perfidious and corrupt. The result was this faux friendly limbo in which no truth claims were possible.

The problem was that without a shared understanding there existed no basis for learning more about what was happening to them. Given this, it was perhaps not surprising to find a new genus of expertise emerging in the Facebook groups, based on the one thing that still carried weight: experience.

"Where can I read more about hell toxin?" someone wrote, in the middle of a thread about air purifiers. "Are there journal articles on that as well?"

"No journal articles," came the response, from the forum moderator, "just reports from experienced mold avoiders."

By which, of course, she meant herself.

The shift seemed to mark a regression to a much more traditional social structure in which the young, or less experienced, deferred to their elders. The power of this seniority was such that one prominent contributor felt that she could only voice dissent privately.

"Because their whole lives are consumed by this 'hell toxin' and 'mystery toxin' theory," she wrote to me, "their cult-like group is very active . . . They are often respected by 'newbies' . . . who don't realize there is no clinical evidence supporting the theory."

It was deflating to arrive at this point, where the very people who repudiated the expertise of mainstream medicine ended up succumbing to their own tawdry version of it. It seemed that there would always be someone to step in to fill the expert role. You could not have knowledge without custodians any more than you could have religion without priests.

It made you appreciate those who did the job well, those who didn't just know the facts but also what they meant in context. Those who combined a depth of knowledge with a breadth of perspective. In a world where 25–30 percent of the population relied on online support groups for solutions to medical conditions or personal problems, and pharmaceutical companies increasingly circumvented doctors, spending vast amounts of ad money (up nearly fivefold since 1997) to directly target consumers, this sort of expert was more essential than ever. And yet increasingly rare.

CHAPTER 9
Paiute Forestry

James was scheduled to speak with Dr. Hope at 3:30 p.m. Currently we were out of cell phone range, having entered a blank zone where the only trace of humanity was the road itself. It wound now between fat towers of rock thrusting up from the rubble. Absent any point of comparison, such features became even harder to interpret. Smaller than mountains yet somehow more dramatic. The land seemed to carry its own idea of how large it was.

It was a surprise, therefore, when we passed a formation on top of which someone had erected a telephone pole, carrying a single wire God knows where, and in the process revealing the formation's dimensions — about the size of an eight-floor apartment building.

A while later we stopped to check out a roadside cabin built by a family of Mormons who, over 130 years earlier, had fled west in search of religious freedom, demonstrating, if nothing else, the necessity of unclaimed land to new ideas. Really weird new ideas, at any rate.

Our journey was but the faintest shadow of theirs, a historical afterimage. The cabin they built was hardly larger than the Rover, yet at one time it had been home to thirteen children. One can only imagine how desperately they must have wanted freedom to endure such conditions.

A bloom of green appeared, and we passed signs for a peach orchard and petroglyphs. Then the green was gone and it was back to pink citadels and rocky escarpments. The land blushed red. But for the peaches, it could have been Mars.

By 3:30 we had arrived in the town of Torrey, population 182. A scraggly black dog stood wide-legged in the back of an old red pickup. Irrigation water jangled along a roadside ditch. It was the kind of town that featured a "trading post" with a tepee out front that had seen better days.

James pulled up outside the Chuckwagon Lodge, the sign for which was adorned with a wagon wheel and enough Tombstone lettering to fill a churchyard. His phone rang and he patched the call through the Rover's speaker system.

I'd been looking forward to the call with Hope. Maybe she would furnish a clearer picture of what post-patriarchal expertise might look like. Could there be such a thing as "matriarchal expertise"?

It was Hope who had originally diagnosed James and referred him to the doctor he was going to see in L.A. Hope, too, had helped frame his paranoia—or prudent precaution, as the case may be. James was in Denver at the time, a year and a half after his initial exposure, and had already washed all of his belongings in vinegar or bleach. Hope told him to throw everything out, even the suitcases. It was the first of many purification rituals. A way to start fresh, like the Mormons. But unlike the Mormons James's exodus never ended. It was as if time had stopped for him, stranding him forever in the opening chapter of the American dream.

Like almost every other EI doc I'd run across, Hope didn't start out in environmental medicine. She practiced family medicine for eight years and would probably still be doing so if she hadn't contracted EI herself. At that point she got board certified in integrative and environmental medicine, and shifted her practice to match.

Depending on how you looked at it, Hope's personal experience with EI either increased her credibility or detracted from it. On the one hand, for instance, anyone complaining of such incomprehensible symptoms was immediately suspect. Then again, that Hope had actually experienced the disease effectively annulled all suspicion of ideological bias—the only other obvious reason someone might have been drawn to the field. Her education and professionalism also worked in her favor. The lack of intellectual rigor that allowed James to nebulize his eyeball, for instance, was presumably less likely in one who had passed her most recent board exam in the top 3 percent.

One might even find in Hope the ideal synthesis of the two modes of expertise that are usually at odds in mainstream medicine: the doctor's and the patient's; knowledge and experience; the world and the self, exterior and interior. Perhaps this was what post-patriarchal expertise looked like. Or perhaps Hope was just as deluded as every other blimp-affrighted sensitive. Doctors, too, could suffer from depression, and any other psychological disorder, so why not delusion as well?

Within the highly variable field of environmental medicine Hope seemed to hold a prominent place, at any rate. She had served as president of the American Academy of Environmental Medicine, an organization known for its outspoken opposition to genetically modified foods, water fluoridation, and Wi-Fi in schools, and regarded either as a den of charlatans and hippie whack-jobs or a collective of forward-thinking researchers and clinicians endeavoring to see past the crippling conservatism of the medical establishment—depending on who you asked. And Hope's pedigree was certainly legit. She had trained with Dr. Michael R. Gray, an Arizona-based doctor of internal and occupational medicine who in his thirty-eight years of experience had seen over sixty thousand patients in his environmental toxicology clinic. Gray, in turn, had trained with Dr. William Rea, the senior practitioner in the field of environmental medicine. And Rea, finally, had trained with the founding father of the field of environmental medicine himself, Dr. Theron Randolph.

For a clinician, Hope's academic output was respectable. She had published three research reviews relating to mold toxicity in peer-reviewed (if marginal) journals. She had also signed her name to two editorials filled with such sentiments as: "Future generations may look back with astonishment and wonder how our culture thought it could stand by and tolerate the poisoning of its people and somehow not anticipate the ravages of widespread disordered biochemistry and ill health"; and "we believe that scientific censorship is dangerous and that the broad spectrum of therapies should be critically assessed equally and evaluated based on scientific merit, not on the medical paradigm 'box' to which they are ascribed."

This second quote appeared in a journal also featuring articles on "ionic footbaths" and "earthing," the practice of walking barefoot outdoors, or otherwise connecting yourself to the earth while indoors (with, say, a copper wire) in order to facilitate the transfer of the earth's electrons into the body—thought to help reduce chronic pain and contribute to sounder sleep. (Our ancient ancestors, the argument went, were far more grounded in the earth than we were, either directly, sole-to-dirt, or else through "perspiration-moistened animal skins.")

Now James was catching Hope up on his various struggles, like the icepick head pain he'd been dealing with since being reexposed to mold toxins several weeks earlier at his West Hollywood apartment.

Hope's responses were supportive but nuanced. I couldn't help feeling that she was trying to bend her talk around James to the writer in the car. It was as if she wanted to signal her full awareness of how implausible the syndrome sounded, and offer a preemptive defense. Chemotherapy patients, she pointed out, for instance, often became hypersensitive to toxins post-treatment.

"People who work with cancer know this," Hope said.

Whatever else sensitives might claim, in other words, the *principle* of hypersensitivity was not outlandish. In fact, Hope argued, lots of people experienced periods of hypersensitivity at some point in their lives. Most cycled out of it. But some did not.

"In spite of their best efforts," Hope said, "it's just really, really tough, and there's something obviously physiological that makes it a lot harder."

It was a canny argument, for it replaced the usual EI contention (that the world was clearly divided between sensitives and normies) with one much easier to defend: that there existed a continuum of sensitivity, of which people like James, Brian, and the Snowflakers represented one extreme. Hope went on to put a little distance between herself and this group, admitting that there was a limit to what could be known for sure about their experience.

"There are people," she said, choosing her words with care, "who have very descriptive . . . identification of certain symptoms with certain types of exposures that generally there's no way to confirm or deny. Y'know, they say my nose itches with Aspergillus. I get that occasionally and I have no way of knowing one way or the other, because there's just no way to know. So it could be completely made up or there could be a correlation."

Among this group of people, Hope went on, there were also those who allowed the syndrome to "dictate their entire existence."

It sounded like a description of James. As if realizing her misstep, Hope added, "And I can talk to you like this because I know you get it."

"Yeah," James said, not really getting it.

"I am in no way saying that environments aren't very significant to health," Hope went on—a curious reassurance for an EI doc to make—"but I am also aware that if you have these hits and then they trigger significant meaning and stress and fear, then that also has an effect. And I know you've been through these cycles, and it's hard, because I know you do well with this, so we can discuss it because you're one of those people who from that point of view handles it pretty well."

It was a remarkably crafted comment, on the one hand telling James outright that his symptoms were largely psychological, and on the other absolving him of the charge. It was also a remarkably politic comment, suggesting someone who was keenly aware of what could be said and what couldn't, and of the inherent difficulties that came with professing expertise on a subject in which expertise was impossible.

So perhaps *this* was the new expertise: one that was earnestly devoted to not saying too much. If nothing else, it highlighted the painful shortfall between what the patient needed and what expertise could offer. In doing so, it made it a bit easier to understand the expert's temptation to overstate her expertise in hopes of giving comfort. Resisting this temptation surely required discipline.

When James floated his idea for a new kind of integrated toxin sensor that could detect chemicals, VOCs, and mold or his idea for an air quality data collection network based on cassettes that you plugged into your air purifier and then mailed to a central lab, à la Netflix, Hope limited herself to the most noncommittal sort of encouragement.

"If you find anything that is statistically significant," she said, "that would add immense value. And it would also add some validation, I guess, for people who are trying to get it acknowledged that something real is happening."

Validation desperately needed by people like Brian and the Snow-flakers, but which Hope herself couldn't offer.

Hope perked up on hearing we were trying to get access to Snowflake.

"I'm dying to hear about it," she said. "It's got this huge mystery to it. We all know *of* Snowflake, but I have yet to actually meet a human who's *been* to Snowflake."

With that the conversation ended and we got back on the road, both of us silently pondering what had been said. In a way, Hope's chariness spoke well of her. She was at least aware of how problematic the field was. She was no hippie whack-job. And even if she was being disingenuous with James, it's not like she had much choice. If a patient couldn't tolerate an insight there was little point in sharing it. Such was the paradoxical nature of expertise: even without the distortions of scale it still gave rise to a breach between itself and its subject.

Yet I was left with the sense that there had been something strangely backward about the conversation. Like somehow it was James's job to unearth the truth about EI, whereas Hope's role was simply to serve as rear echelon support. Like she was headquarters, and he the agent in the field.

When I suggested this to James he was typically unwilling to ascribe blame.

"We who suffer from this are like pioneers of environment illness," he said. "It's still unknown territory to even some of the greatest doctors."

Maybe he was right. Maybe it was wrong to expect the doctors to lead. Maybe it would always have to be the patients.

———

The road continued westward, slowly lifting us onto the Colorado Plateau. Trees began to appear again, piñon and juniper (to which James was violently allergic), and among them black cows tearing at the subalpine grasses.

We were still aiming toward Brian's last known location, but it was getting late in the day and we were beginning to think about where to find lodging. In the meantime we had tuned in to James's favorite podcast, *Philosophize This!*—an episode covering, fittingly, the limits of empiricism.

"On the one hand, eyes are pretty amazing things," the host was saying. "They are these completely unique jelly-filled spheres that absorb the light around you and send signals to your brain which creates this incredible map of the world around you so that you can walk around in it safely. Absolutely incredible. I mean, why do we have eyes if not for this reason?"

"To see fires," James said, suddenly alert.

At first it looked like it might just be a low cloud, but then we crested a rise and could see where it emerged from the pines maybe five miles distant, a gray plume canted east as if from the funnel of a westbound locomotive.

"If those fires are gonna be where we're gonna be tonight, I'll drive and you can sleep," James said grimly, one hand on the breathing mask he kept by the cup holder.

It wouldn't have been the first time James fled a wildfire. It had happened three months ago at his house in Sedona. He felt it before he smelled it, a constriction in the throat, an itch in the eyes. The sun rose that morning with a ghoulish beauty that made James reach for his camera. Online maps showed wildfires to the east kindled by an easterly wind.

"I tried to tough it out," James said. But by the time night fell the smoke had thickened, so he strapped on his oxygen mask and, towing the tank dolly behind him, began packing as fast as he could.

"I went through the house like some kind of cat burglar," he said, "just grabbing stuff and throwing it in bags, and then left to go to Aspen."

It had been a rough summer for the Southwest, fire-wise. By June—two months ago—fire crews were already battling thirty-one wildfires across Arizona and New Mexico, which were seeing the hottest start to summer ever, with temperatures reaching 120 degrees. In less than a week five people had died while hiking. In Phoenix, planes were being rerouted. With 30 million people under heat advisories, dumbfounded meteorologists groped for words. "It's like you took a salt shaker of madness and sprinkled it over the western half of the country," said one.

The incidence of wildfires had actually been on the rise since the mid-1980s. Since then, they'd been occurring nearly four times more often, as well as burning more than six times the land area and lasting almost five times as long. Experts were projecting that wildfire season would soon be year-round. The cause was drier conditions due to the progressively earlier snowmelt—another consequence of climate change.

In this sense the very qualities that made the Southwest so appealing to sensitives could also make it the most hazardous—not least because the methods by which fires were fought were often as noxious as the fires themselves. Take aerially delivered fire retardants. For sensitives, the sight of twelve thousand gallons of bright red chemicals erupting from the belly of a DC-10 was about as close to a vision of hell as you could get.

The predominant fire retardant, in fact—Phos-Chek, a slurried mix of water, dye, and various phosphates and sulfates of ammonia—was for many years made by Monsanto. You couldn't ask for a better case study to illustrate why so many sensitives indulged conspiracy theories. What else but the machinations of chemical companies, for instance, could explain why the use of aerially delivered fire retardants had been trending steadily upward since they were introduced in 1956, despite their well-documented toxicity to aquatic life and disputed overall effectiveness? Two thousand sixteen, in fact, was on track to be among the most profligate years of all time, second only to 1994, which absorbed 41 million gallons of aerially delivered fire retardant—enough to fill sixty-two Olympic swimming pools.

But there were other factors contributing to Phos-Chek's abiding popularity. At the end of the last fire season, for instance, one Forest Service veteran had noted that the pressure to use it often came from local communities who feared the loss of their homes. It was not just the wildfire that the Forest Service was managing but also the fear it inspired.

After all, there's nothing like a dramatic plume of chemicals released from a low-flying plane to assure you that the authorities have the situation well in hand. In this, fire retardant acquired purpose simply by mirroring, in its garish hue and spectacular deployment, the very evil that it fought.

Nor was the Forest Service itself exempt from the fear. Since 1911, after a massive wildfire charred over three million acres in less than forty-eight hours and killed seventy-eight firefighters in a single afternoon, the Service had dismissed light, preventive burning as "Paiute forestry" (a disparaging reference to an Arizona Native American tribe), and instead framed its mission as an epic struggle against the awesome ravages of nature.

"Firefighting is perhaps the nearest thing there is to war," said one forest supervisor at the time.

It was the same metaphor favored by so many high-powered doctors who fancied themselves engaged in a heroic battle against death and disease.

Wildfires, of course, were actually just one component of a far more complex land use issue. But the early Forest Service was invested in a contrary idea—just as so many doctors, chemical companies, and insurance companies who dismissed EI as merely psychogenic were invested in a contrary idea about illness. And the degree of this investment foretold how resistant the ideas were to change.

Consider, for instance, that even though chemical fire retardants have been known to be toxic since at least 1974, the Forest Service was still disputing their toxicity as recently as 2011. That they finally did acknowledge their toxicity was only due to the efforts of a former timber lobbyist named Andy Stahl who, in the wake of a 2004 air drop that killed twenty thousand fish in Oregon, sued the Forest Service on the grounds that they had never conducted an ecological review of fire retardants, and that their use violated the National Environmental Policy Act and the Endangered Species Act.

A year later, a Montana judge agreed with Stahl and ordered the Forest Service to conduct the review. But even then the Forest Service resisted, complying only three years later when the judge threatened to throw the undersecretary of agriculture in jail. Thus it wasn't until 2007 that the Forest Service agreed to expand its testing and monitoring of fire retardants and discontinue the use of retardants that contained ferrocyanide, which becomes toxic when exposed to sunlight.

The Forest Service still refused to change many of their guidelines for using fire retardants, maintaining that, properly deployed, they posed no significant impact on the environment. This despite explicitly contrary findings from the Fish & Wildlife Service.

Stahl sued again in 2008, and in 2010 the judge once again agreed with him, giving the Forest Service until the end of the year to come into compliance. It therefore wasn't until 2011, after fifty-five years of aerial fire retardant use, a seven-year legal battle, and the threat of imprisonment, that the Forest Service finally acknowledged the toxicity of fire retardants. And a year later they *still* hadn't fully addressed safety concerns, with new measures still being implemented—even as new evidence emerged that 50 percent of all aerial fire retardants were still deployed improperly.

Efforts to regulate the carcinogenic and mutagenic fire retardants that suffused our mattresses, furniture, electronics, and just about every other household item had been no more successful. Even when regulators did manage to take an offending chemical off the market, chemical companies could simply tweak a molecule and reintroduce it under a different name. In fact, between 1983 and 2009, global production of fire retardants more than sextupled. In one recent study of California day-care centers, fire retardants were found in 100 percent of dust samples.

The most shocking thing about the Forest Service story was that the agency had no profit motive—yet they still fought tooth and nail for the right to keep dropping millions of gallons of toxic chemicals all over our forests. For they had been sold on the modern promise of chemical convenience, and even when the promise was revealed as a lie they couldn't bring themselves to give it up.

CHAPTER 10

The Story of Color

It was around eight o'clock when we pulled into Panguitch, a small town in southern Utah named for the Paiute word meaning "big fish." Here the road widened from two lanes to six, the way a river widens as the current slows.

It was a sleepy town. To any passerby, in fact, it could almost seem like the last fifty years never happened. It took me a while to figure out why. Mostly it was the fonts. The cars may have changed, and the clothes, but the very words in which the town was written—the store names, the signage—were made with fonts that predated digital typeface and the kerning and serifs that came with it.

The most recent touches to the town—incandescent-bulbed red and teal motel signs with a *Jetsons* flare—seemed to have been added sometime in the 1960s. The rest was brick that carefully guarded its age—though a pamphlet later told us the town was founded in 1864.

We had escaped the wildfire farther north, the road tacking south at just the right moment to deliver us from the smoke. The problem now was finding a place to sleep. All the motels looked mangy in the extreme, worn to the nub by the concussive flow of humanity. Unlike the Element, these were not places that offered the illusion of an immaculate beginning, a chance to feign amnesia and despoil virgin furnishings. These were places that you could not enter without feeling a creepy sense of trespass, and the blurred presence of the thousands who had come before, leaving only the homely husk of dreams, and perhaps a light sprinkling of skin cells.

Anyway, there were no vacancies. It was tourist season as well as fire season.

"This is what I deal with," James said.

We pulled over at a BBQ joint with a sign in the window touting home-made pies, and beside it another wishfully asserting a few old-fashioned certainties: "A good horse, a good gun, a good wife." The sky had ripened like a plum with a half-moon rising in the southwest. The star-rapt silence of the land yawned around us—so unlike the ceaseless fracas of the New England night, abuzz with hidden creatures clamoring to be heard. Waiting for a table, a group of French teenagers took turns doing handstands against a brick wall.

Still no word from Brian, but I did, to my surprise, get a call back from Liz, our Snowflake contact. She agreed to meet the following afternoon but warned against bringing cell phones or any other electronics.

It was a brief if puzzling conversation, and as much as I'd hoped Liz would call, it left me wondering why she'd agreed. As James and I waited for a table I ran through a few brief scenarios in which we were lured into a trap and suspended by our ankles over fuming buckets of off-brand floor cleaner.

James tried to reassure me.

"Before you got here I was a bit concerned," he said. "You coming from the East Coast and . . . but . . . you're fine."

Maybe. But honestly I didn't know what I was. Except tired. And hungry.

We spent most of dinner desultorily discussing washing machines, and then retired to rooms we had eventually found at separate motels—low, one-story deals built around parking lots. The rug of my room looked like one big stain. The walls were hung with pictures of churches, the glass long gone and replaced with cellophane. In the bathroom, mold vined the plastic shower stall, and a chemical reek rose from the bedspread.

To me it was just gross. I would survive. But would James?

▬

It's appealing to believe that there once was a time when such questions didn't need to be asked. A time when you didn't have to be afraid of your vegetables, your pajamas, your soap. Not that it was a safer time, but at least it was easier to know where the threats lay.

How far back would you have to go? And what could explain why everything changed?

You might say that the story began with human vanity. What else? It was human vanity that fueled the rise of King Cotton in the nineteenth century. Between 1849 and 1859 cotton production in the United States increased over 100 percent. Together with advances in textile manufacturing, this surfeit of raw material enabled the middle classes for the first time in history to attire themselves in relatively high-quality ready-to-wear clothes and enjoy the same enhanced self-image that had once been reserved for the elites. This was the era that saw the first large-scale department stores—and fashion magazines featuring haute couture—as the elites sought to redraw the class lines that the sudden availability of quality clothing had blurred.

But the surge in textile production drew attention to a critical bottleneck in the industry: color. Manufacturers were still using dyes that required painstaking labor to produce. One kilogram of red dye, for instance, required 100,000 female cochineal insects, bred in Mexican cacti and harvested by hand. Yellow dye derived from bat guano imported from Peru, and blue from the leaves of a plant called woad, native to Central Asia and once used by the ancient Egyptians to dye the cloth wrappings of mummies. That's how old some of these dyes were. And the production of them simply could not keep pace with the burgeoning demand for swank duds.

This takes us to the first of two unlikely pivots on which so much of the history of chemicals turns. William H. Perkin was born in 1838, the third son of a prosperous builder. The details of his life have been largely dissolved by time, but a few flecks remain, one or two of which are suggestive of the monumental role he would play in transforming the world.

Perkin's grandfather was an alchemist who built a basement laboratory in hopes of transmuting base metal into gold—an ambition that presaged the coup that Perkin, as a precocious teenager, later carried off in a London garret. How well acquainted Perkin was with his grandfather is not known. What is known is that he shared with him an early appreciation of the wondrous nature of reality and the mystery it concealed, and he developed an interest in painting and photography to explore it. At age fourteen he captured one small part of this mystery in a photographic portrait of himself standing somewhat awry against a white wall. The picture reveals a young man who looks much older than his years regarding the camera with sensuous lips and heavy, unfocused eyes.

When he wasn't painting or dabbling in photography Perkin played the violin, and at one point even considered forming a touring string quartet with his three siblings. But he was equally fascinated by the mechanical world gearing up around him, like the steam engines he could see from the top floor of his house on King David Lane, trudging the London and Blackwall railway.

As his interests swayed between the artistic and the scientific, Perkin was eventually drawn to the latter upon witnessing some experiments a friend was conducting with crystals. You can imagine the thrill as he watched the hexagonal forms emerge from nothing, not unlike his own image on the photographic plate, hinting at the mystery that seemed to lurk just beyond the surface of things.

Soon Perkin was skipping lunch to attend a class on chemistry at the City of London School. His teacher at the time suggested he write to Michael Faraday, discoverer of the electromagnetic field, and one of the most influential scientists in history, for permission to attend his celebrated talks at the Royal Institution, and Perkin was thrilled when Faraday wrote back in his own hand, granting his assent. At fourteen, Perkin was the youngest audience member, and possibly the keenest. He could have no way of knowing that the roles would one day be reversed, with Faraday sitting in the audience listening to him.

When he was fifteen Perkin entered the Royal College of Chemistry, very much against the wishes of his father, who, along with pretty much everyone else in the world at that time, saw little practical application for chemistry. The Royal College was a shoestring operation that owed its existence to a German chemist named August Wilhelm von Hofmann, who enlisted a mix of industry figures, as well as the queen's physician, the prince consort, and Faraday himself, as backers. Perkin thrived under Hofmann's tutelage, and within two years became his youngest assistant.

The work at the Royal College was mostly academic in nature but Hofmann would occasionally assign his students specific projects. One of these was to find a synthetic substitute for quinine, the only known treatment at the time for malaria—the bane of the British Empire. Much like dyes, quinine production was limited by its source, the bark of the cinchona tree, which grew almost exclusively in Bolivia and Peru. The British had managed to smuggle cinchona seeds out of South America and were growing them on their colonial plantations in Asia, but supply still fell far short of demand.

Perkin had, by this time, set up his own laboratory on the top floor of his father's house on King David Lane. The laboratory was rudimentary, with neither gas nor running water. During Easter vacation, 1856, Perkin was working on the quinine problem using coal tar, a waste product of coal gas distillation (coal gas was used to light city streets), when his efforts produced a reddish-brown sludge. The result wasn't what he expected, and he tried again. His second effort yielded a darker sludge. Perkin tried submersing the sludge in alcohol and was rewarded by wisps of purple curling toward the beaker's glass walls. Inspired, he applied the mixture to white silk. The result was a rich, fade-resistant purple.

For a painter, all of the world's mystery is contained in color, the way it lies flat on the canvas yet seems to contain fathomless depths. Like music, color transcends its rough origins, aspiring to abstraction. The blue of the sky is the purest blue. The red of the rose is the purest red. Now imagine being able to conjure such colors from nothing. Not to *derive* color, but actually create it. You can imagine Perkin's heart begin to hammer as the purple, like a cautious octopus, unfurled its arms.

At the time, purple dyes were made from lichens, expensive to produce and prone to fade. It was a color with a remarkable history, gleaned, in ancient times, from the glandular mucus of snails. But the secret to this method, yielding a hue called Tyrian purple, had been lost with the fall of Constantinople in 1453. It was said that Tyrian purple was discovered by Hercules himself while walking with his sheepdog on a beach near the city of Tyre. The dog was gnawing on a mouthful of snails when the hero noticed the unusual shade of its drool.

The Tyrian method wasn't any easier than the lichen method—twelve thousand snails were required to color the hem of a single garment, and the stench of the dyeing vats was ghastly. But its difficulty to produce ensured that Tyrian purple could only be afforded by the highest levels of society, the exclusive privilege of popes and kings. It's this that gave rise to the abiding association between purple and royalty.

There was therefore something radically democratic in Perkin's discovery, and his deliberate decision to name it after the ancient regal hue, Tyrian purple. The story only becomes more interesting when we learn Hofmann's reaction to the breakthrough—which was to downplay it and strongly discourage Perkin from pursuing it further. As far as Hofmann was concerned, the purple was an irrelevant accident, and one that could only distract Perkin from more serious work. But Perkin, even at eighteen, somehow had sufficient character to ignore his mentor's opinion, and proceeded to apply for a patent.

Luckily for Perkin, it was around the same time that the empress Eugénie of France, wife of Napoleon III, decided that the color purple complemented her eyes. Thereafter, she began wearing it publicly. Eugénie was a trendsetter. Whatever she wore was soon picked up by the growing number of fashion magazines, many of which featured hand-colored plates. Not long after, Queen Victoria wore purple on the occasion of her daughter's wedding.

Perkin, meantime, was busy consulting with a Scottish dyer to fine-tune the dyeing process. ("I am glad to hear that a rage for your color has set in among that *all powerful* class of the community," the dyer wrote to Perkin. "The Ladies.")

Perkin's synthetic version of the dye hit the market shortly thereafter and was an instant hit. By the spring of 1859 the streets of London and Paris were awash with the color that Perkin had first seen nosing around his beaker. He went on to make tons of money, selling his business at age thirty-six and retiring to pursue pure chemistry.

By then a new era had begun. The era of synthetic color. It was as momentous a breakthrough as the invention of photography by Louis Daguerre, twenty years earlier. What Daguerre invented at that time was not photography, per se, but the synthetic image. Just as Daguerre emancipated the image from that which it represented, so Perkin emancipated color itself from the base and often noxious matter that had always supplied it. Freed from matter, from the snails and the bugs and the guano, color was reborn into the pure abstraction it had always evoked. And yet it could still be manipulated. It remained, somehow, real. For Perkin, it must have felt as if he were dabbling with the mystical quiddity of matter itself.

And just as countless images sprang from Daguerre's discovery, so did countless colors from Perkin's. It began with Verguin's fuchsine, a blazing synthetic red, followed shortly by Nicholson's blue, Martius yellow, and Hofmann's violet. Yes, even the skeptical Hofmann got into the game, eventually.

In 1862 a number of these synthetic colors were featured at the London International Exhibition. "Language, indeed, fails adequately to describe the beauty of these splendid tints," Hofmann effused at the time—and in his wording you can detect a kind of metaphysical awe.

> Conspicuous among them are crimsons of the most gorgeous intensity, purples of more than Tyrian magnificence, and blues ranging from light azure to the deepest cobalt. Contrasted with these are the most delicate roseate hues, shading by imperceptible gradations to the softest tints of violet and mauve.
>
> Nor is the interest exclusively absorbed by the fabrics dyed in these splendid colors; side-by-side with them are exhibited the dyes themselves, some of them beautifully

crystalline, and resembling . . . the sparkling wings of a rose
beetle.

All of these marvelously beautiful colors are derived
by chemical transformations, more marvelous still from the
same prime origin, mainly, from the noisome tar.

From this rainbow our modern world of industrial chemicals emerged.
It began with the shift of synthetic dye manufacturing from England to
Germany in the 1870s—mainly because the British government failed to
support their industry the way the German government did, and because,
unlike the British, the Germans somehow intuited the full implications of
Perkin's breakthrough, which ultimately had less to do with synthesizing
color than demonstrating that chemistry could have profound practical
(and remunerative) applications. The obvious next step was to integrate
scientific research into industry, which the Germans did by developing
the first industrial research laboratories. Henceforth, science and industry
would operate in tandem, the one steering the other.

With this, chemists began to move out of the attics and basements
and into the workplace, leaving behind their dabbling empiricism and
embracing a more rigorous, theory-based methodology. Until this point,
industry had always defined itself by its product. But the rise of industrial
chemistry allowed the Germans to think more in terms of process, thus
enabling the dye companies to start reimagining themselves as chemical
companies. The difference was more than semantic, as the new frame allowed the Germans to redefine what were once dismissed as "byproducts"
as valuable raw materials for future products that had yet to be developed. After all, unlike every other raw material—wood, for instance, or
wool—chemicals were fungible. And in this lay the seeds of the explosive
diversification to come.

It began with pharmaceuticals, which bore a strong chemical resemblance to dyes. In 1898, Adolf Baeyer, who launched his career by synthesizing indigo, introduced aspirin. By 1907, Monsanto was marketing
an anti-fever drug, followed by a laxative two years later. The nascent
photography industry also benefited from Perkin's breakthrough, when

Hermann Wilhelm Vogel noticed that coal tar colors improved the sensitivity of photographic emulsions. In 1909, a veteran of the photography industry, Leo Baekeland, invented Bakelite, the earliest form of plastic, which was used in everything from radio knobs and textile bobbins to the lacquers, enamels, and formaldehyde resins that, a hundred years later, would make sensitives like James reach for their breathing masks.

But if the chemical industry emerged from a desire to look good and attract other humans, what really made it take off, a half century later, was the desire to kill them.

DAY
3

CHAPTER 11

Poison Unleashed

I woke up the next morning feeling hungover, though I hadn't been drinking. And for a moment I found myself in James's shoes. Wondering. What was it? The napalm they used on the bedspread? The grody shower stall? Or just the aftereffects of eleven hours on the road?

Who cares. Take five Advil and let the liver sort it out. In an era drained of compassion and awash in toxic chemicals, toxic news, toxic people, surely the liver displaced the heart as the body's most emblematic organ. The liver, signifying not how we connect with each other but how we preserve ourselves from the ever-gathering yuck.

Outside, the clean air came as a refreshing shock, hammering home the perversity of electing to shelter in a feculent motel. It was early, the sky clear blue, the trees glowing orange from the waist up. Beat-up recliners sat empty outside each room, awaiting the day's load of old men in undershirts. Only the caretaker stirred, pushing a broom around the carport by the reception area.

The effect of mornings such as this was to still you to silence as you realized that nothing you could do could make the morning more perfect, that in fact anything you did would only cheapen it and make it worse. Better to leave that job to someone else. Somehow the caretaker had learned the secret of the morning, passing to and fro with his broom while leaving not the slightest ripple in the splendor.

Inside the fluorescent breakfast nook I served myself bad coffee with powdered creamer and stirred it with one of those inutile red plastic straws. There was no food, per se. Just a heap of cellophaned pastries that telegraphed DO NOT EAT ME. A refrigerator unit knocked and bucked like an old truck. I retreated outside to wait for the fragile day to thicken enough to bear my ugly weight.

James picked me up at eight and we headed downtown for breakfast. He seemed more or less functional, but as we settled into a booth at the local diner and he removed his sunglasses it became clear that his night had not been easy. As soon as he opened the door to his motel room, he said, he could sense the dampness in the carpet. He spent the first half of the night trying to sleep in the Rover, but the front seat only reclined 45 degrees, and in the back he couldn't stretch out his legs.

It was two in the morning when he gave up on the Rover and moved inside. His first thought was to pull the bed to an open window so he could sleep with his head in the night air. It was a brilliant idea and made you wonder why every hotel didn't offer this option: some sort of rubber-lined porthole you could stick your head through to escape the hammering heater and sandpaper air and tune in to the stars and the scents and sounds of the night. Unfortunately for James, his bed was bolted to the floor.

Next, he tried moving the luggage rack and the night table to create a rickety bridge between the window and the bed across which he could lay his inflatable mattress. Obviously, his mattress couldn't be allowed to touch the toxic bedspread, so he encased the mattress within two heavy-duty contractor bags (never leave home without). It seemed a decent solution, but whenever he tried to mount the rickety assembly the nylon of his sleeping bag would start slipping around on the contractor bags.

Two-thirty. Three-thirty. High above, the stars etched their diamond passage on the night's black plate. James tried tucking the sleeping bag inside one of the contractor bags, like a pen inside a shirt pocket. This minimized the sliding but he still woke up an hour later on the floor. Thankfully, the contractor bags did their job and kept the carpet from touching either his sleeping bag or the air mattress. But after waking up six more times James discovered that the air mattress—which he had just purchased—had somehow been punctured. Finally he gave up. For James there was no escape, not even into sleep.

It was a story of compound absurdity, and just hearing it left me vacillating between amusement and despair. Had I actually lived it you might have found me out in the desert, cursing the heavens, or curled up in a trash bag with the Glock.

Yet James didn't appear discouraged. On the contrary he actually seemed upbeat, boyishly enthusing about visiting Snowflake, Kickstartering a tent prototype that you could erect on your roof rack, and maybe moving to Greece and buying an olive farm.

After the Charlie incident, this was the second time I'd seen James exhibit this level of conspicuous tolerance. As breakfast wore on I kept wondering how God would have reacted had Job behaved this way. Probably He would have been pissed. Yet what else could He do to Job but kill him? Slim satisfaction for an angry God.

I shared God's frustration. Surely there was something James wasn't letting on. The match was too perfect: the man who refused to object and the disease that made everything objectionable. I could feel the psychogenic explanation gaining ground.

———

After breakfast we waded across main street to the Rover and James thumbed a button to release the tailgate. I felt a trifle self-conscious standing beside him, in the middle of downtown Panguitch, as he rummaged through the foam-lined suitcase, moving his large hands among the bottles like a concert pianist and popping pills according to a sequence known only to himself.

"This one's the *Limitless* drug," he said, holding one bottle up. "Based on that movie? This guy, his girlfriend breaks up with him because he's not successful. So he takes this pill and it changes his whole world."

The drug was modafinil, an Adderall knockoff. James had learned about it from Dave Asprey, a health guru and self-described "bio-hacker" who built an empire on a coffee brand inspired by Asprey's chance encounter with a cup of yak butter tea in the mountains of Tibet. Or so the legend went.

Like James, Asprey had endured a range of weird ailments growing up. As an adult he suffered from brain fog, and for eight years (until the yak butter discovery), modafinil kept him functional. A brand prescription would set you back $600 per month, James said, but he had learned through his network that you could get an off-brand version from a company based in India for one tenth the price.

I was eager to try it. I'd used speedy drugs before and knew the bracing effect they could have, the way they accelerated cognitive processing and froze out emotion. Perhaps they helped James cope.

Panguitch slid away behind the Rover's tinted windows. We hadn't gone far when James pulled over at a gas station to stock up on 5-hours. On the heels of the modafinil I felt newly interested in every variety of stimulant, so I grabbed a few for myself and popped one in the parking lot. It tasted like something that might leak out of a car battery. Ingesting the stuff, I found myself tuning inward, waiting for the magic to take hold. It was the same with the modafinil. Self-administered substances, regardless of their actual potency, always managed to cast a spell, luring your focus inward and kindling a suspenseful uncertainty about their effects. Illness itself—or the fear of it—provoked a similar response. But there was always a point beyond which the inner landscape faded into darkness, and it became hard to tell what was going on—without instrumentation, anyway.

Not that the impenetrability of the darkness kept you from monitoring it, like a child at night, listening for things that might or might not be there. Was that a bump my heart just gave? An effect of the modafinil, perhaps? "Chest pain," said my phone, listing the possible side effects. "Fast, slow, pounding, or uneven heartbeat. Mood or mental changes. Seeing, hearing, or feeling things that are not there."

Could one of the side effects be noticing nonexistent side effects? I began to feel queasy.

An email had come in from Susan Molloy, the Snowflake gatekeeper, who somehow must have heard we were coming. It was an odd email, pugnacious in tone, caught somewhere between a desire to set the record straight and another to tell us to go to hell.

Mainly she seemed interested in demonstrating that it was pointless to generalize about the EI community in Snowflake. The sensitives who lived there, she said, came in every flavor. Some were Bernie supporters, others supported Trump. Some were "spiritual and cultural fundamentalists," others were liberals, or agnostics who "thrive on the search." Some were blue-collar, others had PhDs. Some had been there over twenty-five years, and many others had moved on. And the range of diagnoses was just as diverse: "asthma, Lyme and/or Tickborne Disease, TILT, CFS, EHS, RSD, CI, MCS, post-polio, lupus, reactive airways, genetic predisposition, toxic encephalopathy, seizure disorders, having crashed in earthmoving equipment while working on a slag heap."

The only real thing they had in common, Molloy said, was that they thrived better in Snowflake than the world beyond, "where we were unable to defend ourselves effectively from fabric softeners, new carpet, roof retarring, cigarettes, crop-dusters, lawn-care chemicals, smart meters, routers, block-by-block cell towers and all the rest."

She did acknowledge the presence of psychological issues, but claimed that these were far more common among those diagnosed in the 1970s and 1980s. At that time, she said, "we were utterly desperate, and we took exceptional and sometimes unfortunately conspicuous survival measures."

What's more, there was, at that time, "no relevant science, treatments, Internet sharing (or Internet at all), legal support, non-toxic construction materials, scent-free products, organic markets, mail-order clothing or supplements, no international statements of support by respected researchers and academics. No footnotes, no references. Nothing but each other."

It was an interesting suggestion, and one that had to be true. That *people* were the resource of last resort. And perhaps that's what made the community, on the face of it, seem so old-fashioned. Because she was right. The Internet had changed how illness was experienced. Now that sensitives could meet up online, meeting in person was no longer necessary.

I took another sip of 5-hour, tongue recoiling from the sharp chemical clutch. Outside it was 90 degrees. The Rover hushed through the blue dazzle, great castles of rock rising to the east. To the west the horizon fell away to a serene distance and then neared again like an inquisitive animal. Eventually the sagebrush and mesquite gave way to Martian plains etched by time and weather. Clouds appeared, and a phalanx of motorcycles roaring north.

Minor muscles that I rarely noticed gradually un-tensed. Thoughts unfurled. It was the kind of feeling you got at sea. The bridge of vision arcing high over the empty leagues but never quite arriving at the farther shore.

No doubt James experienced the landscape differently. For him, even these vacant spaces still held peril. So he remained vigilant, scanning for smoke plumes, coal-fired power plants, industrial farming, high-voltage wires, "and all the rest." Modernity itself was the enemy, having leaked from Perkin's beaker and tainted the entire world.

———

The history of chemicals picks up again in 1868, with the birth of Fritz Haber—the second pivot. Haber left considerably more traces than Perkin. Even so, much remains buried in the riddle of his character.

Haber's mother died in childbirth, and one of his biographers has speculated that his father, Siegfried, never quite forgave him for it. From this initial rift much was to follow—a classic Frankenstein story of scientific arrogance soaked in dire twentieth-century irony.

A dye and pharmaceuticals salesman by trade, Haber's father was a man of circumscribed ambition. As a Jew in late nineteenth-century Germany he was wary of aiming too high and incurring resentment.

Siegfried's relentless caution spawned the opposite in his son, a desire to break loose, to transcend, to—as he once put it in a youthful letter to a friend—"abandon, at all costs, the harbor into which my father has withdrawn himself after arduously weathering the storms of life; to sail out into the limitless ocean of life and future, guided by no other star than by one's own will and striving."

Like Perkin, Haber conducted chemistry experiments in a home laboratory. And his father, too, initially denied his request to study chemistry at university. But where Perkin's father feared that a degree in chemistry could never lead to worldly success, Haber's father, thirty years on, knew that it could, but feared that any success would always be impeded by his son's Jewish heritage.

Like Perkin's father, Siegfried eventually relented. But his fears proved prescient when Haber was unable to find a job upon graduating, nor secure an officer's commission in the military. Defeated and humiliated, Haber had no choice but to take a job working for his father. Tensions between the two were such that within a year and a half Haber's father threw him out. It was around then, at age twenty-four, that Haber got himself baptized, renouncing his Jewish identity and identifying instead with the rising tide of German nationalism, the fervent idealism of which perfectly suited his character.

Haber eventually landed an assistant professorship at the University of Karlsruhe, acquiring, along the way, a reputation as a deeply driven man—power-hungry and easily slighted.

"This otherwise so splendid man has succumbed to personal vanity, which, moreover, is not even of the most tasteful kind . . ." Einstein wrote to his cousin. "Vanity without real self esteem."

In 1906, spurred by the remarks of a rival, Haber began exploring the problem of how to synthesize ammonia, from which fertilizer could be made. At the time, Germany controlled around 90 percent of the world's dye production (the drabness of British military uniforms being but one result), and therefore 90 percent of the valuable by-products. But it was still dependent on Chilean imports for fertilizer, the key ingredient of which—nitrogen—also served as the basis for all modern explosives. A few years later, when the British blockaded German ports during World War I, this dependence would prove a critical vulnerability.

Haber's work caught the attention of BASF—the same chemical company for whom Adolf Baeyer first synthesized indigo twelve years earlier—and they formed a partnership. By 1909 Haber had solved the ammonia problem, a feat which later earned him the Nobel Prize—and, more to the point, freed Germany from dependence on Chile.

And this might have been enough. But with the war Haber transitioned into a new role, advising the government on how to adapt existing ammonia factories to produce nitrates for ammunition, which the German artillery was consuming in unprecedented quantities.

Until this point the fledgling field of chemistry had been seen as a source of beautiful colors and miraculous medicines. It was Haber's genius to recognize that chemicals could also serve Germany's war effort — not just in the production of explosives, but also poison gas.

Chlorine was another by-product of the dye industry. In 1915, Haber supervised its first deployment at Ypres, burying five thousand tanks of the stuff in a long line at a spot called Lover's Knoll, and waiting for the wind. When the wind failed he shifted the canisters ten miles north, releasing them on April 22 at 6 p.m. The gas created a wall fifty feet high and four miles long, marching at one hundred feet per minute toward the Allied lines. The grisly result does not bear description. Among the victims were several Zouave regiments, whose traditional uniforms of baggy red pantaloons, blue jackets, and red fezzes likely got their tint from German dyes.

The Kaiser was thrilled with this novel use of chemicals, and rewarded Haber's boss with a bottle of pink champagne. Haber, too, was thrilled. In his view, chemical weapons set themselves apart from ordinary ordnance — not by their monstrous effect on human biology but rather their unique ability to disrupt a soldier's fundamental powers of threat assessment. As one of Haber's biographers put it, in a description that could easily refer to the plight of sensitives,

> bullets might still kill individual soldiers, but they no longer caused the morale of entire armies to crumble.
>
> Chemicals, on the other hand, represented a many-faceted and ever-changing threat. There could be dozens, or even thousands, of lethal chemicals, each one with its own distinctive smell or taste or color — or none at all — and each one requiring a new kind of gas mask filter. Each new poison thus posed a new lethal threat, and a new psychic challenge to the foe, "unsettling the soul."

Haber's wife, on the other hand, had always loathed the idea of chemical weapons, and shortly after the attack, when Haber returned to Berlin, she borrowed his army pistol, retired to the garden, and put a bullet in her head. She therefore wasn't around to see the introduction of mustard gas ("a fabulous success," Haber wrote), phosgene, and chloropicrin two years later. Haber didn't attend her funeral. Within hours of learning her fate he was off to the Eastern Front to supervise the use of chlorine against the Russians.

Thus Haber escaped the claustrophobic shadow of his father, and achieved the Nietzschean glory he had long sought. Despite his Jewish heritage he was commissioned an officer, appointed the director of a prestigious scientific institute in Berlin, and amassed a fortune through his partnership with BASF.

It's here that the poisonous irony settles in, carried by history's shifting winds. Because it was Haber's own institute which, four years later, developed a cyanide-based pest control agent called Zyklon A. The name may ring a bell, for Zyklon A's younger brother, developed in 1924, would be used to exterminate millions when the Nazi definition of "pest" expanded to include various categories of human. By that time, Haber, his Judaism expungable only by death, had fled to England, a broken man, his fortune lost, his family scattered or dead. He died in exile a year later, having only very late in life come to see technological progress as mere "fire in the hands of small children."

But the chemical industry had by the interwar era developed its own momentum, achieving an importance that rivaled nuclear resources in the wake of World War II. The Treaty of Versailles itself claimed twenty thousand tons of German dyes as Allied war spoils and reserved the right to take an additional 25 percent of future production over the next five years. Meanwhile, industry leaders who had made millions on the war were desperate to diversify lest the entire industry collapse as quickly as it had arisen. This capitalistic imperative fueled the rise of the modern chemical industry in the postwar era. The explosion of new chemical-based products, in other words, had as much to do with the industry's need to justify its continued existence as consumers' need for the products themselves.

The advent of the automobile furnished a hundred new applications for chemicals, from lubricants and antifreeze to radio knobs and artificial leather, and helped the chemical industry achieve solid footing. By the 1930s the chemical industry had switched from coal tar to petroleum as a chemical base, and companies like DuPont were introducing materials like neoprene and nylon, which forever changed the consumer landscape, replacing products like panty hose, lingerie, rubber hoses, and carpet with inorganic fibers. By the middle of the century synthetic chemicals were everywhere, enabling one Time Life writer to observe (with a breathlessness eerily reminiscent of Hofmann's exultant praise of the first synthetic dyes a hundred years earlier) that

> today a stylish woman can go to the theater completely attired in synthetic material, from her acrylic wig to her vinyl shoes. Tomorrow a businessman may be able to leave his all-plastic house in the morning, walk across a lawn of polyethylene grass to his fiber glass car, and drive to work over roads cushioned with synthetic rubber.

It was only a matter of time before synthetic dyes found their way into our food. In the post–World War II era the use of food additives increased dramatically as America's food production and distribution system grew in size and sophistication, and companies began to grasp the full impact of optics on consumer behavior.

It should perhaps not be surprising that food additives served as the original impetus for the field of environmental medicine. It is after all one thing to wear chemicals or even to breathe them, but it's something else to actually eat them.

The connection between food additives and food allergies was first made in the 1940s by Theron Randolph, an allergist by training and environmental medicine's founding father. Randolph's interest in food additives eventually expanded to include pesticides, industrial solvents, and other chemicals, and in 1962—the same year that saw the publication of Rachel Carson's *Silent Spring*—he published *Human Ecology and*

Susceptibility to the Chemical Environment, which outlined the threat that chemicals posed to human health. The book laid the groundwork for a new field Randolph called clinical ecology, which eventually evolved into environmental medicine.

With Randolph's work and the hugely influential *Silent Spring*, the chemical industry faced its first serious public backlash. But the truth is that the ill effects of chemicals had been known from the very beginning. As early as 1862, Hofmann, on analyzing a woman's green ballroom head-dress, found it to be loaded with arsenic. ("It will, I think, be admitted," he wrote, "that the arsenic crowned Queen of the Ball, whirling along in an arsenic-cloud, presents under no circumstances a very attractive object of contemplation.") A German chemist discovered the same thing eight years later in fourteen samples of commercial magenta dye. Letters appeared in local newspapers, complaining that the new dyes caused strange skin irritations. In 1875, the German doctor Richard von Volkmann noticed an increase in scrotal cancer among coal tar distillers, marking the first known instance of an industrial carcinogen.

And yet, by 1884 a London *Times* editorial was already dismissing the call for a return to the organic dyes as impractical. Progress, the paper argued, moved in one direction: forward—the same sentiment that Haber himself would one day use to justify chemical weapons. The mystery of matter had been unlocked, and there was no locking it away again.

It's still amazing to think that a single teenager's tinkering could so profoundly shape the course of history. Fifty years after Perkin's break-through, Hugo Schweitzer, the head of the U.S. subsidary of the Bayer Corporation, went so far as to credit Perkin with the way women smelled. It was Perkin, after all, who once derived coumarin from coal tar, which led to artificial musk, which led to the synthetic production of perfume—which, for sensitives, forever ruined public spaces one hundred years on.

As avatars of the chemical world we live in today, Haber and Perkin make a fitting pair. Between them they neatly bracket the range of human motivation—Perkin the innocent teen, stunned by beauty in a London garret; Haber rather less innocent, orchestrating slaughter on the killing fields of France. If Perkin commercially introduced chemicals into the world, Haber accelerated their impact. Within twenty-four hours of the first use of chlorine at Ypres, the commander of the British forces had wired London demanding access to comparable weaponry. As for Haber's revolutionary synthesis of ammonia, it gave rise to modern fertilizer, which today helps feed billions, but also leaks into the landscape, polluting groundwater, causing algae blooms and acid rain, and disrupting ecosystems everywhere.

Interestingly, Haber, throughout his life, suffered from a mysterious ailment not dissimilar to James's, and regularly sought relief at mountaintop sanatoriums, where the air was clear and the surroundings peaceful and quiet. Doctors could never find a physical cause for his suffering, and eventually labeled it "a nervous disturbance of the pineal gland." Haber's biographers speculate that the cause was psychological, another instance of the "neurasthenia" that was then coming into vogue.

Perkin, too, presents certain mysteries. Why, for starters, didn't he just toss that unpromising sludge, like anyone else would have done? Why didn't he fall prey to disappointment or disgust—the usual human response to failure? And how, for that matter, was he able to dismiss the advice of his teacher and mentor, a giant in the field, and press on with his plans?

One clue might be those lectures by Faraday, which Perkin took in at such an impressionable age. Perkin was seventeen when Faraday spoke at the Royal Institution in London about mental education. The first point made that day was the importance of distinguishing between those mysteries that could be broached and those that could not. In the latter category, Faraday put God. Everything else, he said, was fair game. And in fact the imperviousness of God to human instrumentation could only help refocus our attention on those mysteries that did lie within our power to discern.

In theory, at any rate. For Faraday, like Polanyi, looked skeptically on claims of scientific objectivity. "The force of the temptation which urges us to seek for such evidence and appearances as are in favor of our desires," he wrote,

> and to disregard those which oppose them, is wonderfully
> great. In this respect we are all, more or less, active promoters

of error. In place of practicing wholesome self-abnegation, we ever make the wish the father to the thought: we receive as friendly that which agrees with us, we resist with dislike that which opposes us; whereas the very reverse is required by every dictate of common sense.

Better to recognize the likelihood of bias, Faraday argued, and distrust the prospect of certainty. Unlike Haber, Faraday's watchword was humility, as it later would be for Perkin. Possibly, then, it wasn't teenage arrogance that led him to defy his teacher but the exact opposite: a humble readiness to perceive what was there.

CHAPTER 12

Quackery's Great-Granddad

South, now, through the green tablelands west of Bryce Canyon, and into the dry, alkaline country north of Kanab. The land was opening up again, the bluffs withdrawing and the trees giving way to scrub. Somewhere in there James's leg started to bounce, meaning we had to pull over to liberate some toxins. It was good to be outdoors and upright again, if only briefly. The wind rattled a run of rusted barbed wire and battered the dry grass. Beneath the pale gypsum the earth blushed pink. Something about the waiting stillness immediately made one feel whiskered and alert.

Edward Abbey wrote well about this land, having worked it as a forest ranger in his younger years. His view, unlike mine, a mere passerby, was more tactile. For him it was a land of suffering—choking dust, parched animals, crushing heat. But also beauty. A land stark and unadorned by life but lifelike in its deathly way, the dry hills aglow in the afternoon light, the time-worn rocks as curvaceous as flesh. One could die here and fit right in. This was the other use to which the land might be put. A use to which Brian might have already put it.

Abbey once described finding a dead man here, like me a mere pass-erby who lost his way in the punishing heat and settled against a juniper tree to breathe his last.

"Looking out on this panorama of light, space, rock and silence I am inclined to congratulate the dead man on his choice of jumping off place," Abbey wrote.

> . . . he had good taste. He had good luck—I envy him the manner of his going: to die alone, on rock under sun at the brink of the unknown, like a wolf, like a great bird, seems to me very good fortune indeed. To die in the open, under the sky, far from the insolent interference of doctor and priest.

From the interference of experts, he meant.

Back in the Rover, James put on a Tim Ferriss podcast. Ferriss, the bestselling author of *The 4-Hour Workweek* and *The 4-Hour Body*, among others, was one of James's reliable sources of guidance. Like James, Ferriss, too, was obsessed with instrumentation. A 2011 *New Yorker* profile noted that among the devices in his home could be found "a Zeo headband (to monitor brain waves and sleep cycles), a pulse oximeter (to check the ox-ygen saturation of his blood), a glucometer (to track blood sugar) and a kitchen timer."

With Ferriss, however, the instrument fetish was less a sign of desper-ation than the technocratic monomania of the body hacker trying to eke the last ounce of performance from his machine.

As tempting as it was to dismiss Ferriss as just another huckster (the *New Yorker* profile oozed disdain, as did a *Times* review of his *Body* book, claiming that it read "as if *The New England Journal of Medicine* had been hijacked by the editors of the SkyMall catalog"), he, too, arguably, was advancing a new genre of expertise, one that wrested control from the white-coated old guard and claimed it for the common man. The ethos was pure DIY: the theme was "How to optimize your body" but it might as well have been "How to custom-build a high-powered gaming rig" or "How to turbocharge your car."

The episode James wanted me to hear featured an interview with another of his personal heroes, Wim Hof. Dutch by nationality, Hof was known as "the Ice Man" for his ability to withstand extreme cold. His feats included running marathons in subzero conditions wearing only a pair of shorts; climbing past "the death zone" on Mount Everest, again clad only in shorts; and swimming beneath the ice without goggles or a wetsuit for sixty-six meters. According to the podcast, the swim posed a particular challenge, as Hof lost his sight at 35 meters, and with it his bearing on the exit hole.

"A great experience," Hof enthused.

Hof attributed his unusual cryo-tolerance to certain breathing techniques. He believed that mastering these techniques enabled you to control the autonomic nervous system to a degree far beyond what was generally assumed possible—and maybe even cure cancer. In this, his message echoed Ferriss's: you have far more control than you think—certainly more than the experts would have you believe.

"It is science now," Hof told Ferriss. "And now it needs to get to every person in the world. How to tap into the deeper layers of our physiology—without training years and years and years, and being yogis or super-athletes. My aim is within a couple days we are all able to tap into the deepest layers of our physiology and reset our immune system and bring it under our will."

The appeal was obvious. Like every huckster before him, all the way back to the apostle Paul, what Hof was offering was a cheat, a shortcut, a way to reap maximal benefits with minimal effort. Given the nature of these benefits I could see how his message would resonate with James in particular. In the absence of reliable and accurate threat assessment surely the next best thing was to make threats irrelevant by assuming manual command of one's own immune response.

And yet (leaving aside, for the moment, whether what Hof claimed was even possible), as someone who'd been raised in the Western tradition and seen the destruction that the fixation with control so often wrought, it was hard not to be a tad concerned to see it extended to this new extreme—homeostasis itself. As if we could stroll onto the pitch and do the job better than the biological systems that had evolved over millennia for this singular purpose.

Hof's personal history shed some light on his ambitions. He first developed a taste for the cold in 1979. Twenty at the time, he was sauntering through Amsterdam one day when he passed an ice-covered canal and was taken by the sort of crazy impulse to which only young people who still believe themselves to be immortal succumb. Accordingly, he stripped off his clothes and jumped in.

But it was the second jump that turned him into the Ice Man. This time it was his beloved wife, the mother of his four children, who jumped from the eighth floor of her parents' Pamplona apartment building.

"The inclination I have to train people now is because of my wife's death," Hof later said. Mental illness, he said, can "draw away people's energy. My method can give them back control."

Grief, too, of course, can draw away energy. And what better way to pull oneself together after a devastating loss than to shift the anguish to a physical space where stereotypically masculine coping strategies can be brought to bear?

Wim Hof was a master of "toughing it out," as James would say, and had a number of cold-endurance-related World Records to prove it. It's no wonder James liked him, especially given what a head game EI could become. James had even considered hiring Hof as a personal trainer.

"I love this guy," he said. "He's, like, probably my dad's age . . . Just listening to him the first time, I felt really comfortable with him, as a person."

This, too, I could understand. Unlike most wannabe experts, whose veneer of expertise usually only conveyed how much their product depended on it, Hof didn't take himself too seriously. He pounded beers in his podcast interviews. He began each day with a series of phony kung fu moves. He played guitar, rapped, and liked lounge music. When Ferriss asked him what advice he would offer to his thirty-year-old self he limited himself to saying, "Breathe, motherfucker!"

According to James, Hof's teachings also coincided nicely with brain retraining in that both were focused on reprogramming the primitive brain. On closer inspection, however, this turned out not to be the case. In fact, if anything, the opposite was true. Hof's breathing technique developed as an imitation of the instinctive rapid breathing that came with the sudden exposure to freezing-cold water, a response that Hof compared to the breathing of a woman in labor. Hof's methods, in other words, were based not on transcending or reprogramming the primitive brain but embracing it. Basically, Hof's methods turned you back into an animal.

And this was borne out by what he often described as his greatest triumph, a study published in the *Proceedings of the National Academy of Sciences* that (sort of) proved his claims to be true. The study was carried out by independent researchers at Radboud University Nijmegen Medical Centre in the Netherlands, and entailed injecting twelve volunteers trained by Hof with a bacterial endotoxin to see if they could control their autoimmune response. Normally, endotoxins could be expected to activate the immune system, leading to inflammation, an elevated stress response, and flu-like symptoms. The Radboud study showed that Hof's breathing technique could suppress this activation.

According to the researchers, the mechanism of the autoimmune suppression was a surge of adrenaline triggered by hyperventilation. The adrenaline surge led to a fight/flight response which deprioritized the body's repair and recovery functions in favor of enhanced physical abilities. It was the body's way of saying: Never mind the endotoxin; get your ass clear of that mountain lion. To put this in perspective, the size of the adrenaline surge was, according to the researchers, greater than what was found in a recent study on bungee jumpers.

"I felt like I could conquer the world," one study participant later said. "This feeling, it's insane, it's the same feeling as jumping out of an airplane. The only difference is that this feeling is voluntarily activated by the breathing."

There was a lot to be said for reverting to the primitive—becoming an animal. The animal self offered a kind of mental home plate, and a world in which complexity, interiority, and ambiguity were more easily ignored. In this wonderfully simplified context, the problem—whether endotoxin or profound personal grief—could be reframed in stark physical terms. And for a patient struggling with a difficult illness, maybe this was all that really mattered. Finding, like any fighting general, the optimal ground on which to field your strengths. Considering how different people were, it seemed natural to assume that the ideal ground might not be the same for everyone.

But there was no evidence that Hof's breathing technique did anything to quiet the immune system over the long term. There was even some concern that the immune system might over-activate once the adrenaline surge abated, by way of compensating for the delay.

Even so, the online pile-ons that Hof's wild claims provoked always seemed a bit irrelevant somehow. Hof was only doing what quacks had always done. Selling what people wanted to buy. In the U.S. in particular quackery could boast an unusually robust history, encompassing everything from patent medicines and electric corsets to orgone therapy and energy crystals. The steady persistence of quackery, in fact—despite tremendous advances in medical knowledge—did kind of make you wonder if it wasn't serving some ulterior purpose overlooked by those who were always so eager to dismiss it.

To know for sure you'd have to sift the history and return to quackery's origin. For it did have an origin, in this country, anyway. In fact it began with a single man. It might even be said that this particular gentleman bore as much responsibility for the pervasiveness of quackery in America as Haber and Perkin did for the pervasiveness of chemicals—except in this case the distinction was owed not to what he did discover but rather something he didn't.

━━━

He was no con man, like L. Ron Hubbard, or swindler, like Charles Ponzi. On the contrary, he was the best-known physician of his time, a friend of Thomas Jefferson and a signer of the Declaration of Independence. If not a founding father, Dr. Benjamin Rush was at least a founding cousin, or second cousin. It was Rush who initiated the rapprochement between Thomas Jefferson and John Adams after their friendship fell apart— arguably his greatest feat of healing. Approached to write what later became *Common Sense*, the incendiary pamphlet that helped spur the

American Revolution, he begged off, convinced Thomas Paine to write it, and later gave him the title. He stumped for Washington to be commander in chief, met with him the day before he crossed the Delaware to help buck up his spirits, and, later, infamously betrayed Washington, portraying him as a sloppy campaigner.

Rush's day job, however, was doctoring.

It was an uneasy time for medicine. A pretense of professional expertise had begun to emerge, but little was known about the nature of disease or what caused it. As usually happens, in the absence of real understanding, physicians took what they did understand and applied that—theories of humoral "balance" dating back to Hippocrates and other fancies not far removed from Claudia Miller's pinball analogy. A congruent vagueness characterized how disease was thought to be spread: by "miasma," for instance—bad air arising from swamps, mill ponds, filthy city streets, or rotting vegetable waste. Prescriptions were no less approximate: enough opium "to lie on a pen knife's point," or "a pretty draught" of mercury.

Doctors rarely touched their patients. In practice, they functioned much like psychiatrists, diagnosing disease based on the patient's own descriptions rather than trying to discern it themselves. Disease itself could not be discerned, so there was no point in using instruments. Even when instruments (the stethoscope, the microscope, the thermometer) were eventually introduced, they were little used at first. Instruments represented an insult to the physician's expertise, which was based not on his ability to discern but rather his ability to deduce. Every patient was different, and affected differently—even by the same disease. Only a competent physician could determine what these differences might be.

This was particularly true of fever. Unlike, say, a broken bone, fever presented little evidence as to its cause. This made it a magnet for medical theorizing. Fevers were divided into endless subtypes—based on the season, the rate of pulse, even the frequency of the shivers. There were putrid fevers, nervous fevers, hospital fevers, and jail fevers. Intermittent fevers, remittent fevers, continued fevers, and tertian fevers. And then there was the yellow fever, to which Rush would eventually owe his infamy.

It began in the summer of 1793 in Philadelphia—a trickle of cases to begin with, marked by yellowing skin and black vomit. By August the death toll was mounting—thirteen on the 22nd, seventeen on the 26th, twenty-four on the 29th. By early September nearly five hundred people had died from the disease, which, given the size of Philadelphia at the time, was no small number—the proportional equivalent of 85,000 people in modern-day New York. People began eschewing sidewalks (or the eighteenth-century equivalent thereof) and sticking to the middle of the street to avoid the tainted airs issuing from the homes of the stricken. Handshaking fell out of fashion, and anyone wearing mourning attire was "shunned like a viper."

Rush was the first to identify the disease, having lived through a yellow fever epidemic thirty years earlier. Now at the height of his career, and the ultimate medical authority in Philadelphia (the capital at the time), he quickly took matters in hand, publishing a notice in the local paper advising against ringing church bells and burning fires to ward off the contagion.

Rush believed the disease originated from a heap of rotting coffee on the Philadelphia wharf—largely because the first patients he treated associated their condition with the smell. He also believed the fever to be contagious, and chewed a wad of tobacco during house calls to ensure that he wouldn't swallow his own saliva, which (who knows?) might have been a vector by which the contagion spread. In the murk of the unknowable, precautions of every stripe flourished—cigar smoking, chewing garlic, firing muskets out of windows . . .

Rush began by prescribing a mild course of treatment, including cinchona bark (which contained quinine) and wine. But as the deaths continued his doubts grew and he began ransacking his bookshelves in search of a better approach. Eventually he hit upon a fifty-year-old letter that had been given to him by Benjamin Franklin, to whom the letter was originally addressed. The letter was from a Virginia physician who, in battling something that sounded like yellow fever (but wasn't), claimed to have achieved positive results with highly aggressive purging.

Aggressive treatments were not unknown in Rush's time. Bleeding, for instance, had been a common practice for centuries. It was a relatively easy procedure, and it sometimes worked rather well in subduing symptoms—if only by exhausting the patient. Purging, too, was common in Rush's era—although it's important to note that these practices bore little in common with their modern counterparts. The vomiting, one doctor wrote, was "cyclonic," and diuretics had the effect of "a regular oil-well gusher."

But the aggressive treatment that the Virginia physician encouraged was premised less on its ease or efficacy than a fundamental shift in how medical expertise was viewed. The will of the physician, Franklin's correspondent wrote, must transcend the workings of nature and, more importantly, the suffering of the patient. (An "ill-timed *scrupulousness about the weakness of the body*," he wrote, "is of bad consequence" [emphasis in original].)

For Rush, the letter had the force of revelation.

"Never before did I experience such sublime joy as I now felt in contemplating the success of my remedies," he exulted.

Thereafter, Rush's approach radically altered. For the purge he typically prescribed 15 grains of jalap (derived from the poisonous root of the Mexican ipomoea plant) and 10 grains of calomel (a toxic combination of mercury and chloride and now commonly used as an insecticide)—two or three times the usual dose. Enough, it was said, for a horse. This was given every six hours until the patient achieved "four or five large evacuations."

Rush would then bleed his patients, typically drawing 10 ounces of blood but often more than a pint—sometimes three times a day. This was taking the practice to a new extreme. And as the epidemic wore on, Rush took ever more blood, in one case nearly a gallon in the space of five days. Given the amount of blood the body contains—about five quarts—Rush surely drained many of his patients dry.

Rush's peers denounced him as a "potent quack," a "lunatic," and a "barefaced puff," characterizing his methods as "one of those great discoveries which are made from time to time for the depopulation of the earth." One of his oldest friends described him as "a perfect Sangrado," the lance-happy mountebank in a well-known French novel who "would order blood enough to be drawn to fill Mambrino's helmet [enchanted headgear of a fictional Moorish king], with as little ceremony as a mosquito would fill himself upon your leg."

Yet Rush never doubted the efficacy of his treatments. At the peak of the epidemic he traversed the city poisoning and exsanguinating as many as 150 patients a day, all the while firmly believing himself to be Philadelphia's savior. Not even after four out of six members of his own household died did his conviction waver. "All will end well," he became known for saying.

All did not end well. In the space of three months, yellow fever, together with Rush's brutal methods, killed 15 percent of Philadelphia's population—an episode later referred to as "the most appalling collective disaster that had ever overtaken an American city."

The puzzle of how Rush could fail to notice the ravages his treatments caused has a kind of mesmerizing quality to it. It's not that he was cruel. In fact, he was a great humanitarian. He fought valiantly for prison reform, observing, far ahead of his time, that "the only design of punishment is the reformation of the criminal." He fought equally hard to reform the treatment of the mentally ill, advocating what today would be called occupational therapy at a time when the mad were still treated like animals and exhibited to gawkers for a fee. And he spoke out boldly against the institution of slavery, arguing that if the black man were removed from slavery he would prove the equal of the white man in every respect.

But he also had a weakness for idealism, and as such tended to elevate the principle over the particular. The idealism powered his humanitarianism, but it also made him susceptible to the arguments of the Virginia physician.

As Rush himself wrote, "The conquest of this formidable disease, was not the effect of accident, nor of the application of a single remedy; but, it was the triumph of a principle."

While the shortcomings of Rush's principle-driven medical expertise were obvious to many as the death toll climbed, it took Thomas Jefferson, condemning Rush's style of medicine more generally, to lay them bare. "The adventurous physician," he wrote to a friend, substitutes presumption for knowledge. From the scanty field of what is known, he launches into the boundless region of what is unknown. He establishes for his guide some fanciful theory of corpuscular attraction, of chemical agency, of mechanical powers, of stimuli, of irritability accumulated or exhausted, of depletion by the lancet and repletion by mercury, or some other ingenious dream, which lets him into all nature's secrets at short hand. . . . On the principle which he thus assumes, he forms his table of nosology, arrays his diseases into families, and extends his curative treatment, by analogy, to all the cases he has thus arbitrarily marshalled together. . . . The patient, treated on the fashionable theory, sometimes gets well in spite of the medicine. The medicine, therefore, restored him, and the young doctor receives new courage to proceed in his bold experiments on the lives of his fellow-creatures. I believe we may safely affirm, that the inexperienced and presumptuous band of medical tyros let loose upon the world, destroys more of human life in one year, than all the Robin Hoods, Cartouches, and Macheaths do in a century.

There is of course a self-enlarging effect that comes with adopting a higher principle. An effect that for many may be irresistible, as largeness opens a path to greatness. According to one of Rush's correspondents, a "great man" was what Rush had always longed to become. And in his own eyes this is exactly what he was.

"Let us duly appreciate the difficulties of physicians' studies and labors," Rush once observed, lecturing his audience on the nature of his greatness. "Death presses upon him from numerous quarters; and nothing but the most accumulated vigor of every sense and faculty, exerted with a vigilance that precludes the abstraction of a single thought with the repose of a moment, can ensure him success in his arduous conflict."

Here Rush sounds like the Forest Service firefighters of 1911. Both believed that the response had to be at least as forceful as the challenge.

And in this warlike model both saw themselves as the heroes. Indeed, the aggressive style of medicine that Rush promoted would eventually come to be called (somewhat ironically) "heroic medicine." Practitioners of heroic medicine took pride in remaining unmoved by the agonies their treatments caused (part of why they weren't necessarily pleased by the advent of anesthesia), and would sometimes even brag about how much blood they took. This ability to quantify their interventions was part of the appeal. Heroic medicine offered concrete metrics for an emerging form of expertise that incidentally turned patients into objects.

Having failed the Faraday test—not just lacking humility but actively disavowing it—Rush went on to teach heroic medicine at the university level for thirty-three years, indoctrinating as many as three thousand medical students. This amounted to half of all college-trained doctors in the United States—who then went on to indoctrinate others. Which is how one man established the standard of care for an entire country, single-handedly setting American medicine back (according to one historian) at least two generations.

And not just with regards to yellow fever. Rush also taught, for instance, that childbirth could be eased by draining 30 ounces of blood from the mother at the onset of labor, and then administering one of his cyclonic purges. The practice became common among male midwives for years to come (though not female midwives, who knew better).

And here we get to the responsibility Rush bears for the flourishing of quackery in America, and the tradition that begat guys like Dave Asprey, the guy who built a business empire on yak butter tea. Because even if Rush failed to recognize the folly of his ways, his patients did not, and accordingly sought solutions elsewhere.

Thus the 1800s saw a return to more modest home remedies, together with a new receptivity to what you might call alternative medicine. First among these was an herbal system devised by Samuel Thomson, a backwoods New Hampshire farmer whose mother fell prey to regular doctors ("The doctors . . . gave her disease the name of galloping consumption," Thomson wrote, "which I thought was a very appropriate name; for they are the riders, and their whip is mercury, opium and vitriol, and they galloped her out of the world in about nine weeks"), and later nearly lost his wife to the bloody ministrations of a male midwife. Thomson began his healing career by following around an old woman who knew all the local roots and herbs and what they were good for. A witch, you might say. Eventually he gave up farming to focus entirely on healing, developing his own theory of human health and publishing a book that sold far and wide. The book detailed herbal and dietary solutions that would not be out of place in a contemporary herbal supplement store.

Thomson never aspired to be a "great man." On the contrary, the very notion filled him with repugnance. In the "great man," Thomson saw all the hypocrisy and hubris of the upper class—the traditional roost of experts. "On the lab'rer's money Lawyers feast," he quipped, "Also the Doctor and the Priest; Although their offices are three, They will oppress where'er they be."

His message found an eager audience. At its height, in 1839, Thomson-ianism claimed three million followers—one sixth of the population. Not surprisingly, two thirds of them were women.

After Thomsonianism came homeopathy. Direct forerunner of the alt-medicine industry, homeopathy was the invention of a German doctor, Samuel Hahnemann, who, frustrated by the way drugs were being hap-hazardly prescribed, decided to experiment upon himself to better under-stand their effects. In doing so he discovered that cinchona bark, mainly used to treat malaria, caused the same symptoms in a healthy person that it cured in a sick one. From this he deduced the law of homeopathy: Like cures like. Achieving a cure, in other words, was simply a matter of pre-scribing a drug that provoked in a healthy person the same symptoms one wished to treat.

Hahnemann's second discovery was that heroic doses of medicine were not necessary to achieve the desired results. In fact, in most cases a vanishingly small dose would suffice. So small that the process of diluting the active ingredient probably took more time than acquiring it in the first place. According to Hahnemann, the correct process was as follows: Take one drop and mix it with 99 drops of alcohol. Now take one drop of that and mix it with 99 drops of alcohol. Now do the same thing 28 more times.

When you've finished, take one drop of the final mixture and use it to moisten a clump of sugar the size of a mustard seed. Consume.

Hahnemann was widely ridiculed by the medical establishment— much as sensitives were ridiculed for their fear of trace amounts of toxins. But the ridicule had little effect on the spread of homeopathy. Introduced to the United States by a follower of Hahnemann in 1825, within fifteen years homeopathy began to pose a serious challenge to orthodox medicine as more and more patients were won over by the un-heroicness of the doctors and the painlessness of the treatments. The first nationwide medical organization, in fact, was created not by regular physicians but homeopathists. They even had medical insurance companies on their side, with some offering 10 percent discounts to homeopathy patients during the cholera epidemics of the mid-1800s. Regular physicians, meanwhile, on visiting stricken areas in New York City during one epidemic, were being physically assaulted.

As with Thomsonianism, the success of homeopathy was aided by its rejection of the belittling physiological reductiveness of heroic medicine. In place of this, Hahnemann held that the spirit and the flesh were contiguous, and that disease emerged from a disharmony between man and nature. This far more ennobling view of the human condition quickly found acceptance by Europe's aristocracy. The association with European aristocracy, in turn, increased the appeal of homeopathy among the American aristocracy—or the equivalent, anyway: intellectuals and the upper class. William James was a fan, as were Nathaniel Hawthorne, Harriet Beecher Stowe, and Louisa May Alcott. Once again women were the early adopters, having grown tired not just of heroic midwifery but also watching their children's heads being blistered, their jugulars opened, and their gums scarified with a lance (believed to be effective treatment for teething).

While appealing to different sectors of society, therefore, both Thomsonianism and homeopathy expressed an equal frustration with heroic medicine: the former with its elitism, the latter with its materialism, and both with its ineffectiveness and brutality. For all of these reasons, it just wouldn't do.

It was Rush who created the breach into which all this ebullient quackery bubbled—a veritable Renaissance of it. For after Thomsonian-ism and homeopathy came hydropathy, another import, this time cour-tesy of a Silesian peasant who discovered that applying wet bandages and guzzling cold water fixed him right up after being kicked in the face by a horse. And then the mesmerists, the phrenologists, the Grahamites, each with their own wild theory to tout.

As kooky as these theories sometimes were (Sylvester Graham consid-ered sperm a precious liquid and masturbation a great evil, against which his eponymous crackers somehow constituted the first line of defense), none were as intrusive and dehumanizing as heroic medicine, and all contributed something to our understanding of human health. The Thomsonites recog-nized the importance of eating fresh food. The homeopathists recognized the dangers of overdosing, the importance of "an experimentally derived pharmacopoeia," and spending sufficient time with patients to get a full picture of their health. The hydropathists recognized the importance of ex-ercise and a comfortable convalescence, and also (like the homeopathists and Thomsonites) endeavored to correct for the failure of heroic medicine to address the needs of women, welcoming them as practitioners, offering gentle cures for their children and manuals on how to care for their own bodies. Graham himself, pervy fixation with bodily fluids notwithstanding, is today considered one of the founding fathers of vegetarianism.

Taken together, these movements can be read as something like a collective rebuke to Rush and the medical philosophy that emerged like an ancient curse from Ben Franklin's moldering papers. They also show how what presents as quackery in the moment may actually reveal a cer-tain utility, cyclically highlighting the ways the patient is not being served (else why would quackery appeal?), and more generally the paradoxical tendency of medical expertise to divert attention from the specific pa-tient with which it's purportedly concerned. Every constriction of the di-agnostic aperture creates an opportunity for quackery to dilate it again and broaden the focus—from one small part of the body to the body as a whole, from the body as a whole to the body-mind, from the body-mind to the body-mind-environment, and from there to the realm of the spirit.

Once you start looking for it, this contrapuntal dynamic between establishment medicine and quackery starts cropping up everywhere, an ongoing push-pull of contrary forces across the decades, the one microfocusing attention, formalizing methodologies, and consolidating knowledge, the other macro-focusing attention, subverting methodologies, and democratizing access. You can see it in the rise of homeopathy and later how mainstream doctors regrouped to form the AMA and defend themselves. You can see it in the flourishing of patent medicines, the move toward a more science-based medicine in the late 1800s, and the subsequent degeneration of science-based medicine into paternalism and overspecialization in the century that followed. And you can see it in the recent turn back to homeopathy and integrative medicine, reprising the exact same pattern from two centuries before.

One should, therefore, in theory anyway, be able to assess at any given moment how narrow and myopic medical expertise has become by considering the kind of quackery it has engendered. The renaissance of quackery in the first half of the nineteenth century is a case in point. The renaissance of quackery in the last twenty years (Reiki, crystal healing, herbal remedies, juice cleanses, aromatherapy, craniosacral therapy, reflexology, etc.) is another. Both eras were characterized by an unusual narrowing of expertise, which in turn prompted a lush flowering of quackery. When it came to weighing the dispute between sensitives and mainstream medicine, then, it again became a question of not whether skepticism was warranted but of whom? Sensitives for indulging in quackery? Or mainstream medical experts for succumbing to myopia?

Ironically, Rush himself recognized the utility of quackery. Thus he advised his students to always listen to old wives' tales, because you could never tell when they might come in handy. As he informed the American Philosophical Society in 1786, it was "from the inventions and temerity of quacks that physicians have derived some of their most active and useful medicines."

When establishment medicine got a bit too rigorous for its own good, in other words, quackery kept a back door open for disruptive truths. For as much as we like to think of science progressing by dint of sober analysis and orderly method, many of the greatest scientific advances emerged from the sheerest humbug. It was mesmerism, for instance, that set the stage for the insights of Charcot, Breuer, Janet, and Freud, just as phrenology (in pioneering the notion of localized brain function) set the stage for neurology—and alchemy, for that matter, set the stage for synthetic color.

How then to structure one's own ignorance? How to position yourself on the frontier of the unknown? How to look at the dark? With the narrow rigor of the scientist? Or the billowing humbug of the quack?

CHAPTER 13
Toughing It Out

And so into the beachy wastes of Arizona, following the road south until the old red bones of the earth rose again from the sand, and there was something sort of wrong about the way we ripped their stillness from them as we passed. Ever onward, past nowhere motels with weed-choked parking lots, a tent church hawking healing to the Navajo Nation, a two-pump gas station hawking clean restrooms, pouring ourselves down the dark cone of road, speed and temperature neck and neck. Once, topping an incline, we beheld a biblical valley that had a foreclosed feel to it, as if the sea that once lived there just woke up one day and decided to move on. Farther on, a great starship of crashed rock rose against the horizon, the Rover buzzing along like a black fly beside it. Time meant nothing here. Barely a minute ago Fritz Haber, too, was rolling through this region on a nationwide tour to meet other scientists and take the measure of American industry. One wonders what he made of all the emptiness. Perhaps the indifference of the rock spoke to him somehow, validated his Nietzschean bent. There was much to admire in time's monstrous crush.

Flagstaff gagged us as we came in. The stench was extraordinary, a heavy, turd-meaty reek that coiled into your sinuses and settled. A quick search revealed the source: a nearby Purina dog food plant, one of the widely recognized downsides to life in Flagstaff. "On its worse days, I imagine it's exactly what a zombie apocalypse would smell like," one resident commented on a local Internet forum.

The plant had been in the news recently as complaints had finally forced Purina to study how the stench might be mitigated. A brief peek under the hood of the dog food industry revealed a four-legged version of the same toxic perils plaguing the human food industry—except worse. In 2007, thousands of cats and dogs died after eating pet food contaminated with a compound used in the production of Formica, laminate flooring, and dry-erase boards. Since then—and despite tightening regulations—analyses of dog food had detected unacceptably high levels of a variety of other contaminants, including lead, arsenic, mercury, BPA, cadmium, even mycotoxins. More recently, the FDA discovered traces of pentobarbital, indicating that euthanized cattle were used as ingredients. It didn't get any prettier from there.

And just as Whole Foods had thrived in an era of chemical contamination and food fear, so, too, had various high-end dog food companies, like Blue Buffalo, which for three times the price of the regular slop offered antioxidant-rich premium dog food entirely free of animal by-products and made from only "the finest natural ingredients." Except that even Blue Buffalo was proving unworthy of trust, having been forced to admit in court (after being sued for false advertising by Purina) that their dog food contained poultry by-products like "egg shell, raw feather and leg scale."

In a sense, dogs were the perfect customers. They gobbled up everything without complaint and when poisoned they rolled over and died. Pet activists were always trying to change this and get the dog food companies to treat pets more like humans. But in the meantime human food companies were always looking for new ways to treat humans like dogs. Given the market capitalization of the human food companies, it seemed more likely that they would succeed than the pet activists.

Pressing onward through the stench we soon arrived at the Flagstaff Whole Foods, our lunch appetites only moderately tempered. James knew every Whole Foods between Aspen and L.A. His travels in fact could probably be plotted as a jagged line connecting one store to another, all over the Southwest. The Whole Foods network made it easier for James to keep weight on—a challenge complicated by his elaborate regime of supplements, but more generally attributable to an abiding distrust of food itself. At his lowest he weighed 150—compared to a high of 200.

As with most things, however, Brian had it worse. Limited to four or five foods, his all-time low was 112 pounds. This in a guy over six feet tall.

"Watermelon and carrots," he posted at one point. "That and ghee keep me alive."

Having been a pathologically picky eater as a child, this fear of food was not altogether foreign to me. For a long time my favorite food remained the only one I knew for sure I could trust: me. Or my thumb, anyway. Even the few foods I later favored shared a certain thumby quality—what people today call "umami." But really it's "thumb." Thus I subsisted on Cheerios, macaroni and cheese, and Chef Boyardee ravioli—all foods high in thumb.

It's interesting now to try to recapture this old fear, perhaps my closest reference point for life as James and Brian experienced it. What exactly had I been afraid of as a child? Adults seemed to think that the problem was taste. But the real problem preceded taste. The real problem was trust.

It was as if I'd suffered from an exaggerated sense of my own purity, in that virtually anything could despoil it. It was as if I had endured something traumatic—the trauma that comes with noticing your place on the food chain: well below monsters and anything else more than three feet tall—and remained trapped in a state of needless vigilance. The irony being that my fear blinded me to where the real threats lay. (A close look at the ingredients of Chef Boyardee ravioli revealed all sorts of worrying stuff. Nonorganic, pesticide-encrusted tomatoes, feedlot beef, high-fructose corn syrup, processed soy meat supplement, MSG, and, most impenetrably, *flavorings*.)

You could sense the same sort of hypervigilance in perusing Bruce

McCreary's EI home-building protocol. The endlessly detailed specifications put you in the mind of a child fussily preparing for sleep: arranging all the stuffed animals, dictating how far the blinds should be lowered, which lights are left on, how much the door is left open . . . Both the child and McCreary were building fortresses. As was the child who refused to eat. By keeping out foreign matter, they preserved the fragile integrity of their being. Which was another way of saying that they couldn't imagine admitting whatever lurked beyond those walls without compromising that integrity, and being consumed from within by the ick.

It was this hypervigilant behavior that led some critics to dismiss EI as nothing more than a baroque form of PTSD, much as the Academy of Medicine had dismissed Gulf War Illness. Pall, the NO/ONOO-cycle guy, had already cited a biochemical connection between the two phenomena. And many other researchers had noted a causal link between early-life stress and the likelihood of a dysregulated immune response in later adulthood.

On reflection, in fact, one couldn't help being struck by the parallels between the two phenomena. Both entailed a compulsive search for a means to reconcile divergent realities (in PTSD research the term was "apophenia"—looking for patterns where no patterns existed). Both had been dismissed as syndromes that only afflicted women (it was only the Vietnam War that gave PTSD clinical legitimacy—if men were experiencing it, it had to be real). Both were characterized by a sense of being cut off from the world, or marooned—although with EI the marooning was rather more literal. Both were treated with some variant of brain retraining. And, most tellingly, both, studies had shown, were predicted by a history of child abuse (a fact, incidentally, which was sometimes cited to explain why EI disproportionately affected women, who were more likely to suffer sexual abuse in childhood).

Brian had already admitted to having had a difficult childhood.

"My father was pretty verbally abusive to me," he'd told me. "And he really verbally abused my mom. And my sister. It really affected her."

Brian's father was a factory worker, a heavy smoker and drinker who, according to Brian, was just passing along the abuse he got from his dad as a kid. He died when Brian was still in his twenties, but for Brian the reckoning continued. Nor was he afraid to acknowledge it.

"If I'm being 100 percent honest I know that has a lot to do with how I've processed stress," he'd told me. "You always have those voices in the back of the head."

The PTSD conjecture also conveniently explained how Brian, despite having recoiled from all physical human contact, was nonetheless able to thrive online, in a disembodied world where physical violence was not possible.

James, too, had experienced a certain amount of adversity growing up. He'd always had allergy issues, and often had trouble breathing. When he was five his tonsils were removed. Then his adenoids. Then came the weekly allergy shots. When he was eight he almost died after being stung by a swarm of yellow jackets. He could still remember the feeling of his throat closing up, and lying on a gurney as a nurse pulled off his socks and yellow jackets came swarming out.

To these adversities his father always had the same response, the response that James, even today, regularly recited to himself: *Tough it out.* Not the most sympathetic response, perhaps, but still a far cry from the kind of abuse Brian had to deal with.

The worst thing that happened to James growing up was probably his parents' divorce. James was ten at the time, and identified with both parents. It couldn't have been easy watching his family torn asunder. Still, it wasn't clear that any of this rose to the level of actual trauma—unless there was more to the story.

———

Having stocked up on lamb bars, coconut chips, and kale flakes at the Flagstaff Whole Foods, we shot back out into the desert. Halfway to Winslow the land reddened again and we began to see signs directing us to an AM channel that played a looped recording about a meteor crater dating back sixteen thousand years—the final touch on an already otherworldly landscape. The tone of the recording evoked an ultra-square high school science teacher from maybe seventy years ago. But what was seventy years out here? What was sixteen thousand? The draw of the crater was not its age or four-thousand-foot diameter but rather the way these measurements called its own meaning—and by extension everything else—into question.

The land here had a forlorn quality, scrubby and unowned, and little was lost in driving fast. It was hard to believe that anyone would willingly live in such a place. "Like ground zero at Hiroshima," one source had told me, describing Snowflake. Some years ago, evidently, a pair of farmers had affixed a heavy chain between two tractors and dragged it across the land, leveling all the piñon trees. An unlikely Eden.

And despite the desolation, perils abounded.

"That's a coal-fired power plant," James said, indicating a grove of smokestacks off the starboard side.

I'd read somewhere about coal-fired power plants. They were huge contributors to air pollution, which caused 6.5 million premature deaths every year. They were also the biggest emitters of airborne mercury and arsenic—45,676 pounds of mercury in 2014, 77,108 pounds of arsenic—when even homeopathic concentrations constituted a serious threat to human health. At levels of 50 parts per billion, for instance, water laced with arsenic caused cancer in one out of one hundred people. And just one seventieth of a teaspoon of mercury in a twenty-five-acre lake was sufficient to make the fish unsafe to eat.

A few miles farther on James gripped the wheel tighter as what looked like smoke from another forest fire came into view. But the smoke soon proved to be dust blowing off some dirt piles. But was it just dirt? Or was it mine tailings?

Mining operations posed another hazard. Once in Idaho Falls James ended up in the ER after he felt like his throat was closing up. It could have been anything. There was a lot of heavy industry in the area, including a phosphate mine that had recently received national coverage after petitioning the EPA to loosen selenium emissions standards in a report that included a picture of a two-headed fish.

Not to worry, the mining company said. We did the research and the risk is negligible. Then someone ordered a review of the research and it turned out to be not quite so negligible as the company claimed.

This was the road we all drove, sensitive and normie alike. This is what it felt like. Not knowing. Hurtling blindly through the landscape, evil seeping in at the margin. Who could remember ever not feeling this way? That the true risk could never be known? That the odds were beyond reckoning?

We accommodated it. We were nothing if not adaptive. Within the last few decades we had even come to accept "at riskness" as what one writer called a "fixed attribute of the individual, like the size of your feet or hands." Everyone was at risk for something. If not EI then cancer, if not cancer then autism, if not autism then plummeting sperm counts, rising waters, errant financial algorithms, school shootings, Russian hacking. Whether we wanted to "take a risk" was no longer the issue. The risk was there regardless, the unavoidable cost of modern convenience. In the course of adapting to this risk we gradually came to accept a new level of passivity. There was nothing we could do. We belonged now to risk the way people once belonged to their own country. We were citizens of it.

CHAPTER 14

Hyenas, Nukes, and Tornadoes

This fundamental shift in the nature of risk wasn't exactly a new idea. It was first expressed (in English, anyway) in 1992 by a German sociologist, Ulrich Beck, who, unhappy with the vagueness of "postmodernity," coined the term "Risk Society." Society, Beck argued, was about the production and distribution of goods. Risk Society was about the production and distribution of bads. In recent years these bads had undergone a significant change. Where once they were local, now they were global; where once they were easy to quantify, now they eluded summary; where once they were circumscribed, now they were open-ended. Beck liked to cite the example of Chernobyl, a global catastrophe the full scope of which could never be fully known. "The injured of Chernobyl," he wrote, "are today, years after the catastrophe, not even born."

What's more, as we were seeing in the case of climate change, no institution could be trusted to safeguard against these threats. Our own regulatory agencies couldn't even safeguard against arsenic in our baby food.

Beck's scathing term for this was "organized irresponsibility"—a "logic of institutionalized non-coping." Only pretense preserved the appearance of normalcy, and at times it seemed like the entire system would collapse beneath the weight of this pretense. For not only were our institutions incapable of thwarting this new class of threats, they were equally incapable of compensating for the damage should those threats occur.

The 9/11 attacks, for instance, caused over $43 billion in insured losses, an amount that was met by a thundering "Nope" from insurance companies, which soon ceased offering terrorism insurance. Eventually the federal government had to step in, but even government coverage was capped at $100 billion—when it had been estimated that a mere 10-kiloton bomb detonated in Grand Central Terminal on a typical workday would cause over a trillion in damage (not to mention the deaths of half a million people)—an event to which security experts several years ago had attributed a 20 percent likelihood over the next ten years. The average person was therefore left with no way to insulate himself against the gravest threats of our era. It was as if we had all suddenly found ourselves living in homes without roofs. It was as if the very notion of roofs had disappeared.

Weirdly, we were good with this. More or less. Probably because we had forgotten how different things used to be. This was the curious thing about reality. Unlike every other consumer item there was only one that was readily available at any given moment. This made it hard to bear in mind all the others. The ones that might be and the ones that had been. Life after all was not always so uncertain—not, at least, in quite the same way it is today. There was a time, for instance, rather a long time ago, when uncertainty was presumed to be beyond our control. When bad things happened—floods, famine, highway robbery—the cause was attributed to powers higher than ourselves: the disfavor of the gods, fate, the one-eyed crone down the road.

So what happened? Was there one specific moment when the risk landscape shifted? Better yet, one specific person on whom we could heap all the blame? Or did it happen gradually, blurring the footprints of any potential bad guys? As with the history of chemicals, there was a story to be told.

This one begins with the signature of a Grecian merchant at the bottom of a contract stipulating that he would pay 680 gold florins to insure the shipment of ten bales of woolens from Pisa to Sicily. It was February 20, 1343, and on that day (a Wednesday) the modern insurance industry was born. The term "risk" itself, in fact, is a derivation of the Italian *risico*, from the Latin *resecum*, a figurative term meaning cliffs, like the cliffs of Homer's Scylla, against which the ship of any seafaring hero (or uninsured merchant) could easily be smashed.

From Italy the new practice of marine insurance quickly spread throughout Europe, spurring further commercial risk taking and igniting fresh interest in controlling the future—the one body of uncertainty larger than the sea. By the 1500s, insurance had grown into a legitimate business with London as its epicenter. There, merchants and sailors would meet at dockside coffeehouses to exchange news. One of these coffeehouses, Lloyd's, grew to become something like a global information processing center, and later developed into one of the modern world's insurance giants.

With the success of marine insurance, and the new appetite for security that came with the rise of the bourgeoisie, new categories of insurance soon became available—life insurance first, in 1706, followed by every other kind as risk was gradually transformed into a commodity. Even then, however, risk was still calculated by rule of thumb—except in the case of life insurance. For unlike the life insurance companies, which based their calculations on mortality statistics, companies selling non-life insurance had no access to solid data.

It was this that made life insurance risk calculators the first "experts" in the modern sense, for they were able to profess knowledge of specific incidents while having no personal knowledge of those incidents whatsoever. At the time, most expertise was of a decidedly different quality— what one scholar has called "vernacular expertise"—that is, expertise based on personal experience and transmitted hierarchically from vets to noobs. A riverboat pilot, for instance, learned to negotiate the risks of a river not in riverboat school but by standing for years beside the captain.

There was a lot to be said for such a system: it intimately connected experts with the object of their expertise; it cozily integrated the authority of the expertise within the local social hierarchy; and it privileged a comfortable past (from which all lessons arose) over an anxious, indeterminate future. It had, in other words, breadth as well as depth.

Which was probably another reason (along with the lack of solid data, and the mathematical prowess to make sense of it) why it took so long for disembodied expertise to displace the rule of thumb. But with the advent of the industrial age and the new category of man-made risks that went with it, change was inevitable. The cholera and yellow fever epidemics produced by the crowded and unsanitary conditions in cities eventually gave rise to modern epidemiology—the men in white coats. The explosion-prone steam engines that powered the riverboats (as well as factories and rail transport) eventually gave rise to engine inspectors. Meanwhile the growth of the rail system introduced a new kind of risk, the kind that arose from complex systems. The world's first head-on train collision occurred in 1837 between a lumber train and a passenger train a few miles outside of Suffolk, Virginia, killing three and maiming thirty. In the decades that followed, safety engineers began to appear.

The emergence of this new breed of professional risk managers marked a shift in how reality was perceived, removing a whole class of threats from the care of the average citizen and (conveniently) leaving her free to focus on the growing spread of consumer delights that would define the coming century. It also gave the state a new role as employer of professional risk managers—and a manager of risk itself. Railroad accidents, for instance, eventually gave rise to state railroad commissions, just as the cholera and yellow fever epidemics led to new municipal sanitation laws and unreliable steam engines led to the first federal laws regulating technology. Meanwhile, Otto von Bismarck, chancellor of Germany, hedging against the broader risk of economic disruption and political instability, was introducing the first state-run, state-mandated social insurance—a precursor of Social Security.

One pleasing quality of the new breed of modern risks was that (compared to the typhoons and hyena attacks of yore) they proved relatively easy to track. The carnage of World War I, for instance, yielded troves of useful data. One Zurich doctor, Heinrich Zangger, usually remembered as a close friend of Einstein, took advantage of this data to pioneer a statistical analysis of the effects of Haber's chlorine gas on the human body, thus paving the way for probabilistic toxicology. "Risk and its scientific control must remain in step," he observed, lest heedless progress lead to "chronic, permanent, unseen, degenerative damage to mankind."

This optimistic view of science as risk's keeper was typical of the first half of the twentieth century. New tools were being brought to bear with, for instance, the publication of the first mathematical model for non-life insurance in 1930, which in turn encouraged the growth of a new class of expert actuaries. Meanwhile, as infrastructure veined through the landscape, an electrical engineer named George Campbell began urging Bell Telephone to use probability theory to manage the uncertainty in their phone system. Bell agreed, and soon integrated a math lab into their corporate portfolio, much as German industry had integrated research labs forty years earlier.

The 1930s brought another round of state intervention, except this time the risks weren't cholera and steam engines but chemical exposures in the workplace, pesticide poisonings, and toxic food additives. It was around this time that toxicologists began to establish specific thresholds called "maximum allowable concentrations" (MACs) under which certain substances were believed to be safe. The MACs represented the logical conclusion of a notion first put forward by Philippus Aureolus Theophrastus Bombastus von Hohenheim, aka Paracelsus, a sixteenth-century Swiss-German physician-alchemist. Recognized today for introducing chemistry to medicine, Paracelsus justified the medical use of noxious chemicals with the argument that "solely the dose determines that a thing is not a poison." In this he was the first to articulate the twinned logic of homeopathy, which held that the same substance that sickens us may also serve as a cure. (Five hundred years later the story is still told of how Paracelsus saved a plague-stricken village in northern Italy with poop pills — an iota of a plague victim's feces snuggled in a wad of bread.)

The advent of thresholds marked the beginning of a new, precautionary approach to risk (at least as far as toxic substances were concerned) wherein the goal was the complete elimination of potential harm. In 1937 MACs were embraced by industrial medicine in the U.S., and a year later the Food, Drug, and Cosmetic Act was passed, updating the Pure Food and Drug Act of 1906 by requiring manufacturers to prove that drugs and food additives were safe before bringing them to market. In the years that followed, MACs spread to other fields and began being used to regulate pollution in the air, soil, and water. In the consumer market their influence could be seen in the appearance of new benchmarks like "permissible dosage" and "acceptable daily intake." And so for a decade or two it seemed like Zangger had nothing to worry about.

But once we'd found a way to contain the existing threats we couldn't help inventing new ones. A bewildering array of synthetic and petrochemicals, for starters. Beginning in the 1940s a deluge of such substances poured onto the market. Even then, experts were warning that some of these substances were dangerous at any level, and yet the vast majority were never regulated, for even if the nation's still embryonic regulatory agencies did have the capacity to restrict the more dangerous substances, it's by no means clear that they would want to, because this was the flesh of modernity itself, all the stuff we could not do without: asphalt, solvents, light bulb sockets, celluloid, washing machine impellers, spacesuit helmet visors, Day-Glo paint. It wasn't until 1958, with the passage of the Delaney Amendment, barring any measurable trace of synthetic carcinogens from foods, that the idea of "unsafe at any level" began to be accepted.

Then came the threat of nukes—and not just that one might land on you, atomizing you and all your photo albums, along with 65,000 of your neighbors. By the time Little Boy and Fat Man were dropped, the builders of the bombs had learned that even minuscule amounts of plutonium could cause cancer. Thus by 1948 the National Committee on Radiation Protection and Measurements had kissed threshold theory goodbye and endorsed the view that in all cases exposure should be "as low as possible."

Nukes (a shining example of technology creating at least as many risks as it negated) actually had a marvelously clarifying effect on the way we thought about and managed risk. After all, the quality control measures that might suffice for railroads and phone systems clearly would not do when it came to ICBMs with one hundred times the destructive power of Little Boy and the potential to trigger thermonuclear war. In 1961, the folks at Bell were still the reigning experts on system-level probability analysis, so when the time came to evaluate the reliability of ICBM launch control systems, they got the job. It was there that one of history's overlooked geniuses, H. A. Watson, about whom absolutely nothing is known today, not even her or his first name, came up with fault tree analysis. Which constituted a pretty sweet step forward in the risk assessment department.

With fault tree analysis, you identified potential bad outcomes (e.g., ICBM topples on launchpad, detonates) and then diagrammed every link in the causal chain necessary for that bad outcome to occur. This allowed engineers to assign discrete probabilities for each link and identify the ones that were unacceptably weak. The result was a meticulous dissection of causality itself, as well as a limited expression of the Laplacian notion that complete knowledge of current conditions conferred the power of prophecy. It was a way to drop a bubble dome over the greatest risk in human history—the "worst-case scenario"—and inspect it with a ballpoint pen.

Between the deluge of synthetic chemicals, the dangers of radiation, and the menace of the worst-case scenario, threshold theory began to seem almost quaint, and the impression only deepened in the 1960s with the advent of nuclear power. By then the insurance industry had already made clear that they would not fully cover damages resulting from a reactor meltdown, thus prompting the federal government to step in with the Price-Anderson Act in 1957. The act was supposed to be temporary, until such time as the nuclear power industry could demonstrate a reliable safety record. But given the risk calculus of worst-case scenarios, that time has still not arrived, and after repeated renewals the act is currently set to expire in 2025.

By 1962, with the publication of Rachel Carson's *Silent Spring* and the rise of the anti-nuke and environmental movements, it was clear that we were living with a new kind of risk, one that was not clear and exogenous like the USSR, but murky and endogenous, the product of our own activities. The borders of individual nations could not contain these new threats any more than the threats could be contained by individual generations.

This is what had become invisible to us today, the way we'd accommodated these new threats and come to accept that they would always be with us—not because something could always go wrong but because even when everything went right the risk still lingered. It was the dawn of Beck's Risk Society, and it left us in a bind. Because if the threat could not be eliminated then the choice was either to admit our helplessness and vulnerability, repent of modernity and start working on some new vision of the future to replace "progress"—or else find some way to justify the threat's existence.

The latter path was easier, and so we took it. We had made the transition to a mass consumption society and there was no way back, just as there was no way back from synthetic color. Already, products were just as good as they needed to be and no better. An acceptable level of failure was built into the system. Which brings us to one of the greatest inventions of the 1970s, high fives and hacky sacks notwithstanding: *acceptable risk.* Henceforth, risk would no longer be considered in isolation but rather in relation to what it bought. For every modern advantage there existed an acceptable number of dead people to pay for it. It became a question of costs and benefits. As Robert N. Proctor writes in his book *Cancer Wars*, suddenly risk "is not something to be feared but to be embraced: risk is inevitable, necessary, even desirable in so far as it is 'potential profit.' . . . This is the entrepreneurial ethic writ large." It was beginning to smell a lot like Reagan.

There was an unpleasant cynicism to the new risk calculus, but its acceptance was made easier by the release of the Rasmussen Report in 1975. The Rasmussen Report was a reactor safety study that had been ordered three years earlier by the Atomic Energy Commission in anticipation of the renewal of the Price-Anderson Act. Its main contribution was to familiarize the public with the concept of probability. Nuclear meltdowns might sound frightening, for instance, but in reality the risk of dying from one was six orders of magnitude less than dying in a car accident, four orders of magnitude less than being killed by a falling object, and three orders of magnitude less than being killed in a tornado. The Rasmussen Report was later found to have serious methodological problems and was disavowed several years after its release by the Nuclear Regulatory Commission, a mere three months before the meltdown at Three Mile Island, but by then we'd already grown rather attached to this new mode of thinking. Theoretically, any risk could be lived with so long as it was marginally less than slipping on a bar of soap.

What constituted an acceptable level of risk, however, remained debatable. How many dead people were you willing to tolerate for any given modern amenity? A complicated question. And one that could only be answered by . . . *experts.*

Thankfully, the risk expert business was thriving. By 1957 the insurance industry had birthed the first professional organization dedicated to actuarial risk assessment for non-life insurance (a foretaste of the probabilistic thinking that would later come to dominate evidence-based medicine). In the 1960s and 1970s an entire cadre of risk experts emerged to cope with the worst-case scenarios introduced by the nuclear industry, where a full-on, level 3 probability risk assessment (to gauge the impact of a meltdown on the surrounding populace) could require more than one hundred person-years of analysis. With the meltdown of the Three Mile Island reactor in 1979 probabilistic risk assessment (PRA) became a part of the federal plant licensing process. A year later saw the establishment of the first professional society of risk analysis, and the following year they began publishing the burgeoning field's first journal. At that time the society boasted three hundred members. Six years later the number had quintupled.

But now a new problem began to emerge, because when it came to what constituted "acceptable risk" even the experts disagreed. As Proctor notes, as late as 1983 80 percent of industry scientists still believed in carcinogenic thresholds—compared to 37 percent of scientists employed by regulatory agencies. This was the point where expertise ended and politics began. Like so much else, it all depended on where you were coming from.

The transformation of risk from a purely scientific to a quasi-political concept began to be formalized in 1977, when Shell, Procter & Gamble, and Monsanto, together with 130 other chemical companies, assembled an advocacy group called the American Industrial Health Council to lobby Congress about occupational health policy and the regulation of carcinogens. The AIHC urged Congress to distinguish between scientific and political considerations when weighing the risk of chemical exposure, and in 1980 explicitly suggested that $500,000 be devoted to study, as one scholar put it, "alternative means to elaborate scientific judgments on the quantitative aspects of human risk, especially in the case of chronic diseases"—in other words, ways to put a positive spin on diseases contracted at work.

The money was soon forthcoming, and three years later, at a gala D.C. dinner attended by twenty-four players from the chemical industry, the National Research Council released what came to be known as the "Red Book," which laid out a new approach for risk-related policymaking in which scientific analysis was only one part. The second part addressed other factors, like financial considerations, political considerations, and ethics. The Red Book came as a fulfillment of Reagan's famous 1981 executive order, which stated that, henceforth, "Regulatory action shall not be undertaken unless the potential benefits to society for the regulation outweigh the potential costs," and throughout the 1980s and 1990s it was adopted as the new international standard.

The Red Book provided the framework for the controversy that followed, much of which took shape around the terminology—like "hazard" and "exposure"—because it was the product of these two values that ultimately determined the "dose," and therefore the risk. Depending on which value received greater weight, different assessments of the same risk could vary dramatically. Industry folks liked to claim that the size of the hazard didn't matter if the exposure was zero. Sure, nuclear reactors were hazardous, but so long as they never melted down, what was there to worry about?

Environmentalists and public safety advocates, naturally, felt otherwise. Because what exactly was meant by "exposure"? Exposure of whom? For a long time, for instance, the benchmark for the risk of chemical exposure was the Vitruvian man—a 155-pound working male. But the body of a 155-pound working male was actually rather different from that of a 110-pound woman—and depending on the epigenetics even similar body types could react to the same exposure differently. What's more, studies were beginning to show that, not only was the fetus unusually sensitive to chemical exposure during certain periods of development but the effects of the exposure sometimes couldn't be seen for decades.

The same sort of debate took shape around what was meant by "cost" and "benefit." Industry advocates tended to emphasize the short-term costs to themselves (of pollution regulations, for instance) and the short-term benefits to society (of cheap doohickeys), whereas environmentalists and public safety advocates emphasized the long-term costs to society and the short-term benefits to industry. Not all of these values were equally easy to calculate, and since environmentalists and public safety advocates tended to focus on those that weren't and industry advocates tended to focus on those that were, the latter always had the advantage.

A similar asynchrony could be seen in the values aligned with each of these positions. As Proctor notes, in the view of environmentalists and public safety advocates the individual body was complex, fragile, passive, and in need of constant protection, while industry advocates viewed the individual body as active, virile, resilient, more than capable of detoxing whatever chemicals it encountered and repairing genetic damage as it occurred. Put more simply, one side thought society was strong and the body was weak and the other thought the opposite. The disconnect went as deep as evolution itself as, in the environmental view, synthetic chemicals were a radical departure from the threats our bodies had evolved to handle, whereas in the industry view they were mere variations on threats that had been around forever.

Ultimately, industry always came out ahead, not just because industry had more money, and its stated costs and benefits were more calculable, but because its message was positive and paired better with what we wished were true. With the deregulation that followed, the job of managing risk became increasingly decentralized. Like never before, risk management became a matter of "personal responsibility." With this, the door was opened to an entirely new class of products expressly designed to help the consumer manage risk: personal fallout shelters, 9-volt smoke detectors, retractable seat belts, residential alarm systems, virus protection software, school shooter panic buttons, as well as a deluge of new insurance products and a cacophony of professional expertise on how to avoid heart attacks and raise healthy children. The age of disaster capitalism was upon us, spelling an end to what one scholar has called "the last strongholds of vernacular risk culture: the house, yard and nursery." Henceforth you couldn't even be an expert in your own home. This was great for, say, parenting magazines. Maybe not so great for the mental health of parents.

The final stroke arrived with the disgrace of the professional expertise for which we'd traded our more homely, vernacular ways of knowing, as reports began to appear claiming that large numbers of scientific studies—even in big-time journals like *Science* and *Nature*—could not be replicated. Meanwhile the reversal of previous findings about what was safe and what wasn't had become routine. The day after you finally threw out all your BPA water bottles, for instance, and replaced them with new, safer, more expensive BPA-free models, a study would appear claiming that BPA substitutes were equally dangerous as BPA. All of which left you wondering whether the entire thing wasn't some kind of scam cooked up by the water bottle people.

Exhaustion set in. Defeat. "Living in a world risk society," Beck wrote, "means living with ineradicable non-knowing." It was as if we had circled back to a more primitive world in which the reasons for things could not be known—except this time lacking even the meager consolation of blaming our misfortunes on the disfavor of the gods.

Would we have chosen this? If the full picture had been made abundantly clear of all that we'd have to sacrifice for modernity's delights? Or were we just being led along by our appetites?

Probably we were just being led along by our appetites. And probably we would have chosen this path regardless. History, it seemed, was just a long, haranguing exposition of human nature's follies. And history would carry on so long as human nature remained what it was.

In the meantime we could still try to console ourselves with the wily promise of science and technology, in the hopes they would, per Zangger, keep us one step ahead of the risks that they themselves created.

Thus we had people like Eric P. Loewen, PhD, president of the American Nuclear Society, who, five months after the Fukushima nuclear disaster, would confidently assure his colleagues that "the next generation of nuclear power plants will be even more stable—and we do not need to stop licensing in the U.S. or construction elsewhere to 'go figure it out.'"

Besides, what was the alternative? Simply stop moving forward? Stop making stuff? Stop trying to make things better? If the future was not to be progress then what was it to be?

To this no one had a good answer. "Progress" was the only answer we had.

CHAPTER 15
Snowflake

At the northern edge of Snowflake James and I were welcomed by a sign informing us of the town's elevation (5,635 feet) and date of founding (1878). Trees began to appear again — multiple varieties — even grass, and baseball fields. A few blocks south we hooked a left and the greenery began to unravel again, the tidy businesses and ballfields giving way to car repair shops lost amidst acres of baking wreckage, curtained houses crouched low like reptiles with their elbows out, and nameless ranches marked by white uprights and a crossbeam over the cattle guard offering a gateway into further nothingness.

A few miles on we turned again and soon arrived at the road we were looking for. A jumble of mailboxes met us at the end of it, as if the postman dared proceed no further. Easing the Rover forward, James commented absently on the road surface, which was "not exactly heavy black petroleum asphalt." And so onward, past an "open range" sign, and another saying "no uninvited," the pavement becoming a dirt road, the dirt road becoming a thistly track leading straight to a distant batch of clouds on the horizon. There seemed to be nothing here at all. Just emptiness, stillness. Silence.

The Rover ambled along, gravel popping, and eventually we arrived at a T junction and guessed left. A one-story house with a green metal roof stood amidst several trees. We parked in front of a low brick wall and James stayed with the Rover as I uncertainly approached the front door.

Several minutes passed as I stood waiting for a response to my knock, but the house was as quiet as a tomb. Then came the jostle of a chain and a bolt sliding back and the door opened. Through the screen I saw a tall, emaciated man in his sixties or seventies emerge from the darkness. Eyes sunken, mouth a mere crack, mottled skin tight over his skull. He looked upon me with a distant expression that made me wonder what world I had summoned him from, and whether the usual social customs still applied. After a brief, murmured exchange it became clear that the house was the wrong one, and that we should have turned right where we turned left. I retreated hastily to the safety of the Rover, leaving the specter to subside into the shadows from which he had emerged.

James and I agreed that this was very spooky, and it occurred to me that this might be unnerving for him, to see others of his tribe farther down the road to dissolution. It was an uncomfortable position for him, unable to fully identify with either the Snowflakers or with me. As usual he was stranded somewhere between.

We soon located the correct address a quarter mile or so in the other direction, a similarly constructed one-story house on a slight rise at the end of a long dirt driveway. Two metal sheds stood perhaps twenty-five yards north of the house, and beyond those a tent had been erected beneath a white canopy. Beyond the tent was an older-model Airstream trailer—known among sensitives as a relatively safe brand. Two other small houses were also in view, each perhaps a quarter mile distant. The rest was open range.

James parked the Rover at the end of the driveway and, like encyclopedia salesmen, we began the awkward walk up. An indent in the northwest corner of the house, perhaps nine feet by six, served as a little porch overlooking the range, and we could discern several figures seated there. The palpable antipathy I sensed from these figures even from afar said more about the gulf between them and the mainstream for which I guess

I served as envoy than the remote, godforsaken country to which they had been effectively exiled. Being consigned to this role helped me see how offensive it was to sensitives, and how ignorant I was of the experience of those I presumed to regard as my hosts.

They were in their mid- to late forties, a man and a woman, regarding us silently as we approached. The man wore flip-flops, cargo shorts, a gray tee, and a baseball cap. He had a pleasant, intelligent face, thinly bearded, and a civil, academic air. The woman, who I gathered was Liz, wore a loose spaghetti strap dress and sunglasses on a string around her neck. Her brown hair was home cut in a practical bob, her wide face and blunt nose reddened by the sun, with a paler raccoon mask around her eyes.

The man arose, introduced himself as Scott, and offered us chairs. Liz did not get up. She remained stoutly where she was, watching as Scott did the work of smoothing our transition into their world.

The chairs were arranged around a small table. Behind us, steel wire shelves held a cradled phone and Tupperware filled with light bulbs. A door led into a darkened kitchen. We took a moment to adjust, regauging the range and the way its aspect changed when viewed from a fixed location. It felt different to be in their company. Scott sat with his legs stretched out and his hands linked behind his head. Even Liz despite her air of distrust lounged languorously in her chair. There was a comfortable feeling of physical indolence, a looseness in the limbs. From a little red cooler Scott produced several bottles of Izze — a brand of carbonated juice that boasted "no preservatives, no caffeine, and no evil sciency chemical concoctions."

In describing what followed I cannot be too specific, as it went on for a long time, and for most of it I wasn't feeling very good, the queasiness I'd been feeling earlier having worsened into nausea, probably because of all the Addies, 5-hours, and kale chips I'd ingested.

Scott started the questioning, cordially, as if we were just making small talk, but soon Liz, evidently losing patience, interrupted with blunter questions of her own.

Basically I was being asked to explain myself, what I wanted and why. And so I did the best I could, which wasn't very good at all because the truth is that at the time I don't think I fully knew what I wanted, or why. I had been drawn into this story by a muddled combination of factors. A certain native contrarianism, for starters, and impatience with heedless authority. A growing alarm at what felt like an ever-deepening ambient uncertainty. A peaking frustration with corporate dominance and government incompetence that didn't offer any particular outlet. And an obscure identification with a man whose own solution had been to drop out entirely and withdraw to a remote wilderness.

Meandering among these points, I searched for something concrete to offer. After each sally there would come these long, uncomfortable silences interrupted only by the hooting of the wind in the Izze bottles and a cow's distant, lonely complaint. It was a silence with which they, unlike me, were very comfortable. They reclined against it, whereas I felt it arrayed against me, a kind of aural stare. The space, too. The heat, the arcing crickets. The half-buried cinder blocks with grass sprouting from the holes. Scott's silent trembling water glass, like the needle of some instrument designed to detect lies, half-truths, truths contaminated by vanity and confusion. All of which I felt full of, rising in my throat with the nausea and shoved down again by the pillowy suffocation of the silence.

At some point another woman materialized to watch the strange foreigner babble, an older woman in an orange shirt and hiking boots and cargo pants tucked into her socks. She looked familiar for some reason, like an image from a Dorothea Lange photograph, sun browned and dour with the hard times of a lifetime stitched into the little wrinkles around her mouth. From beneath a sun-faded baseball cap she watched me with bright desert eyes.

Several times Liz got up and withdrew behind the house to confer with the old woman, and then they'd return and sit as before in the crushing stillness, the old woman's eyes on me like a rifle on a prairie dog. Their distrust seemed impenetrable, and part of me felt, sensing now the level of anguish that must be behind it, that it would be imprudent for them to trust a fellow such as me. That it was in fact almost violent of me to expect from them such generosity. It was as if I'd arrived from some grotesquely rich country at the doorstep of its impoverished enemy, begging to defect.

Who would believe it? And yet the longer I talked, the more I felt like theirs was the side I wanted to be on. Out here in the desert with the outcasts who seemed to share a deep, unspoken understanding with each other. And not back wherever I came from with the others. The cruel ones, the stupid ones. The arrogant ones who thought there was nothing they did not know.

And then suddenly it was over.

"Okay," Liz said, and looked at me expectantly, awaiting my first question.

And for a moment I didn't know what to do.

———

The stories were familiar. Scott was twenty-seven in 1998, holding down a job as a research assistant while working toward his doctorate in psychology. He had no prior medical issues and no particular preoccupation with chemicals or toxins.

"Before I got sick I had no idea what sort of toxic world I was living in," he said. "You see a bug, you grab the Raid, that's what you do."

Then his office got sprayed for termites. Scott had used an exterminator before when his apartment was overrun by spiders, but this time they sprayed Dursban, an older organophosphate pesticide which only two years later was (in a typically flimsy regulatory move) banned for indoor use. (The EPA later took steps under the Obama administration to strengthen the ban against Dursban but these efforts would soon be reversed by Trump's pick to lead the EPA.)

Scott got sick immediately. He saw a number of doctors to little avail. In time he got a new job developing a computer-based training program for Home Depot's corporate office in Atlanta. As his sensitivities continued to spread he did his best to ignore them but the more he ignored them the more he stumbled into new exposures, further aggravating his sensitivities.

At this point Scott still had no idea what was wrong with him.

"Allergy was the only frame of reference I had," he said.

Two years after the initial exposure it occurred to him to do a web search for "chemical sensitivity," which is when he finally realized what was going on.

Meanwhile the Atlanta pollution was becoming harder to tolerate. Home Depot let him telecommute for several years, but then one day he came home to find that his apartment had been painted because the apartment manager had given the painting crew the wrong apartment number.

Scott spent the next three years trying to find housing he could tolerate. After looking at over a hundred prospects he finally settled on a place. It was no better than any of the others but it had a covered porch.

"I actually spent a year living outside on that covered porch and just using the house to shower and cook," he said.

When that year was up he bought a different house but that, too, proved intolerable. So he spent another year on a porch. Winter was particularly difficult, as the cold made it hard to use his laptop.

"I remember my fingers swelled up to twice their normal size and I could barely move them. I was, like, what the hell. But I absolutely did everything I could to keep working because, y'know, part of it was I liked my job. And pride certainly was a part of it, like, I'm not going to let this thing beat me."

In 2007, nine years after the Dursban exposure, Scott finally surrendered and moved to Arizona.

"They paved the road right in front of my house. I just said fuck it and left."

He was living in Dolan Springs, about an hour southeast of Vegas, when his landlady asked him to show a house to a prospective tenant— Liz. They dated for a year but eventually subsided into a companionable friendship.

"Just to kinda clarify that," Scott said.

There was a fastidious self-awareness to the way Scott told his story. He was always hedging, amending, ending his remarks with a little proviso or addendum, as if to anticipate judgment.

"I'd rather be out there kicking ass, not hanging out here," he'd say, "not that there's anything wrong with that." Or, "Not that there's anything particularly terrible about EI." Or, "I don't want to come across like I'm some kind of victim . . ."

It was painful to watch these helpless exertions, like a squirming fly in a web. I felt bad for him. I would have liked to tell him that I didn't need to be reassured.

Except that it probably wasn't me Scott was trying to reassure so much as himself. For eighteen years he had been doing this, struggling to maintain a coherent sense of himself as the disease inexorably edged him to the margin.

"This is far from the worst thing that happens to people," he would say. Reasonably, intelligently. "I have a lot to be incredibly thankful for. And I am. It's not the way I saw my life going, but . . . You just, you make the best of it, because what else can you do?"

Clearly he had a great stake in believing himself to be this reasonable, intelligent person. But illness has a way of splitting us from ourselves. Exposing the lie. Laughing at those parts of us we work so hard to make convincing.

Even if Scott ever did manage to fully surrender to his fate, that surrender would never be accepted. This was the great cruelty of the disease. The way it never ended.

The wind nudged scents from the field. Manure, creosote, heated sage. Twenty yards away a jackrabbit went still amidst the quivering grasses. Much farther out a great sledge of clouds tipped to the horizon, so it looked as though the sky and the horizon had somehow gotten misaligned. Eventually the old woman sat down, cradling in her lap a mason jar of iced tea. But those osprey eyes never blinked, and suddenly I realized where I'd seen her before—in a picture accompanying one of those sardonic magazine stories that had made the entire EI community tremble with rage.

Beside her fierce vigilance Liz looked particularly loose-limbed and casual in her striped sundress. Having decided to talk she now began to lay out her story. She started her career at Intel, she said, hired right out of business school and quickly advancing from purchasing to senior buyer. From there she moved on to Motorola where she helped launch their Iridium Satellite system. Her last job was as a marketing manager for Rain

Bird, the world's largest lawn irrigation company. It was a high-powered lifestyle with sixty-hour workweeks and lots of travel and Liz loved every minute of it.

Liz's telling of the story was refreshingly matter-of-fact, and entirely devoid of the reflections and justifications that had marked Scott's narrative. Indeed, she had already snapped at Scott several times over his endless hedging.

Liz was a doer. And whether she was doing work or doing sickness, she did it the same way. Straightforwardly, without guilt or hesitation. Unlike Scott she had not been split by the disease. And there was a certain triumph in this. A charisma. Which is probably why the others seemed to organize themselves around her. Scott as her handmaiden, reading her news from the magazines she couldn't handle or relaying stories he'd heard on the radio. The old woman as her fierce protector, even accompanying her when, complaining of feeling "a little off," Liz retreated fifteen feet with her chair into the dirt and scrub. (It was probably something on our clothes, she said.)

As a gentle, somewhat neurotic intellectual Scott was a familiar figure to me, but there was something magnificent about Liz: the odd combination of her blunt Midwestern pragmatism and her almost sensual comfort in her own physical body; the strength of her character and her extreme vulnerability; the concreteness of her thinking and the sunburn of her skin. It was a kind of magnificence that would have been far less visible in, say, Wisconsin. Unlike Scott, forever fiddling and hyper-attuned to appearances, Liz simply was who she was. There could never be any doubt about her dimensions. She took up exactly as much space as she required—not out of selfishness but because it would never occur to her to waste energy worrying about things she couldn't control. All of which had probably served her well as a high-powered executive.

That came to an end in 2001 after a surgery to remove a uterine fibroid.

"The surgery lasted six hours instead of three because the doctor wanted to peel it out of my navel rather than do a C-section," Liz said bitterly. "Later she told me, 'I didn't want you to have a scar.'"

After the surgery she started getting migraines and body pain. A doctor diagnosed her with Lyme disease and then dispatched her to Dallas to see Dr. Rea, the reigning authority on all things EI, who confirmed that she was both chemically and electromagnetically sensitive.

Liz didn't know what triggered it. It might have been the post-surgery Cipro. ("I've been told that can be a hard drug to detox.") But even before the surgery she'd been feeling unusually fatigued, which she attributed to a series of car accidents in the late 1990s. But the problem may have begun even before that, with the moldy house she grew up in, or the allergies she had as a kid. Or earlier still, with her genetic makeup. Testing had revealed an anomaly associated with a decreased ability to process toxins.

Naturally, Liz preferred to think of her EI in purely physical terms. Traumatic brain injury, for instance, was the term she identified with, and one that accommodated her history of car accidents. But it also implicitly acknowledged the possibility of a psychological component to the trauma, and some sort of post-traumatic stress response.

Her official diagnosis, from a doctor she saw to apply for Social Security, was "organic brain disorder"—a catch-all term meant to signify mental disorders with biological origins. But Liz confessed that she had no idea how either of these terms squared with the toxicity model preferred by the EI docs.

"They're in parallel universes," she said.

"Is it possible that the etiology differs?" Scott said. "I mean, that the sensitivity is a symptom, not necessarily a discrete etiology? So, y'know, maybe there are three different underlying forms of it that have not yet been fully articulated."

After offering this theory Scott immediately backtracked, saying that he'd been "trained for five years in grad school in the scientific method" and really believed in "not getting ahead of the research." But my sense was that he mainly said this by way of telegraphing his understanding of the difference between knowledge and speculation. It was important in other words that we didn't think him a kook.

It was mostly lip service. Because Snowflake was a place where such distinctions ceased to matter. Liz, for instance, was relying on the advice of an alternative medicine doctor who had her on a diet of bone broth and sardines. Like most EI docs, including the one James was on the way to see in L.A., Liz's doctor would probably be considered a quack in the eyes of mainstream medicine. But this wasn't the point. As Liz put it, if a package of sardines could cure her migraine, it hardly mattered who prescribed it.

━━━

"Do you guys like sardines?"

Nausea notwithstanding, it would have been gauche to decline.

The light was fading, the tyrannous sky giving way to something more somber as colossal reaches of whale-blue cloud blanketed the range. The tensions had eased, replaced by a tentative camaraderie as we sat together around the little table. A white metal washbasin held organic grapes and several tins of salmon and sardines—an odd little picnic on the range.

Life in Snowflake wasn't all bad, Scott was saying. The housing was decent. With a twenty-acre minimum lot size there was no shortage of room, and at $300 per acre the prices couldn't be beat. Unlike in Georgia, where he was from, you didn't have to worry that your neighbors were using ChemLawn (though Liz's Airstream did sometimes sprout mold when the rain blew horizontal). People knew each other and helped each other, celebrated birthdays together. And the town offered various sensitive-friendly services: a seamstress who didn't use fragrances, an organic grocer who delivered, a barber who made house calls. The people were friendly, too. Very accepting.

The talk rolled on at a more relaxed pace, James offering his thoughts on salt caves, cryotherapy, and cold baths, Scott and Liz listening politely. The disease, I was coming to see, meant something very different to each of them, challenged them in different ways. Scott the gentle intellectual, Liz the downed alpha, and the old woman, the hardened warrior, I guess. As Susan Molloy, the Snowflake spokesperson, had said, there was no way to generalize. The diversity of their characters and temperaments made it even more difficult to imagine how EI could be purely psychological. It's not as if they were all anxious, or shirkers, or dying for attention. In fact they were about as different as people could be.

When we finally took our leave and were rolling through the hulking dark, a few images kept returning. The night compressing the last ruby light to the horizon, the meager dinner picked from a tin, the way the old woman wordlessly shifted her chair to sit beside Liz in the hot sun, and last of all the tender way Scott, at one point, reached across to touch Liz's shoulder. Alone on the range, they had become for each other what had been denied them. They had become caretakers in a world where care was lacking.

Maybe this is why the old woman finally spoke up in the end. Because enough care had been shown. We'd been talking about where they would go if they could go anywhere. Scott said he would visit his parents. Or return to his office and go back to work. Or maybe just go to a movie. Life with EI, he said, was "sort of a prison where you're locked out rather than locked in."

Liz said she'd probably go to Tulsa to visit family and friends. Her sister's kids were particularly important to her, since the disease had kept her from marrying and having kids of her own.

It was then in the companionable dark that the old woman had spoken up, her voice a low croak. It turned out this parched, sun-baked enigma had spent her childhood boating on Lake Michigan. And that's where she longed to return.

"It's in your blood," Liz suggested.

"Yeah, it's in your blood," the dark figure said.

——

In the dark of the car I settled back into James's presence. Our budding acquaintanceship had taken root in the climate-controlled cabin, and it had been jarring to experience him in the presence of others. I had tried to meet James where he was, but I could sense Scott and Liz wondering about him, and withdrawing ever so slightly to make room for his various untempered enthusiasms.

This was the difference between James and the Snowflakers. They were physically disconnected from the world. But James was disconnected in other ways. Socially, it seemed all he could do was make exclamations—about lamb bars, salt caves, or whatever—and wait for others to pick up on them.

A certain fluency was absent. He reminded me of the Tin Man in the forest, struggling to move. So much about him was fitful and awkward. The way he walked, talked, breathed. Even sweating, he'd said, didn't come naturally to him. He could bake in a sauna for half an hour and emerge bone dry.

This was James's handicap: nothing about him flowed—least of all his anger, his emotions. Which perhaps explained his abiding fascination with Charlie, the Jungian shadow.

The only time James truly flowed was behind the wheel of the Rover.

Uncomplaining, he drove. Despite looking more than ever like a hungover tennis pro, he was, he said, ready to drive all night. But in the end we agreed to save our strength, and set a course for Show Low, a thirty-minute drive.

DAY
4

CHAPTER 16

The Figure in the Static

If Snowflake got its name from two men who worked well together "Show Low" was the product of two men who didn't. According to the hotel breakfast attendant it began with a feud between ranchers who decided that the town wasn't big enough for both of them. They agreed to resolve the dispute by drawing cards. Whoever showed the lowest card could stay. The other had to go. The first guy drew a three and figured he had it made. The other guy drew a deuce. And somehow the town took its name from the story's moral.

Show Low was also the site of one of the largest forest fires in the history of the Southwest. In 2002 the Rodeo-Chediski fire destroyed over 468,000 acres and led to the evacuation of thirty thousand people. It was started at one end by an out-of-work firefighter who hoped to get hired to put it out and at the other by a stranded hiker trying to signal a helicopter. The fire burned so fast and hot that it created its own weather. In a single day a record 86,000 gallons of fire retardant were dropped on the blaze, covering some houses. The brand was Fire-Trol, like Monsanto's Phos-Chek mostly a mix of water, dye, nitrogen, and phosphorus—essentially the same fertilizer product that Haber unleashed in 1909. And indeed measurements taken six weeks later at the Lake Roosevelt reservoir (the water supply for Phoenix) showed nitrogen and phosphorus levels that were, respectively, 22 and 390 times higher than EPA standards.

With all the fires and chemicals—not to mention high-voltage power lines and coal-fired power plants—Arizona hardly seemed like the safest place in the world. But at least the hotels were good. So far, anyway. After disabling the window lock with his Leatherman and tying back the drapes James managed to get a few hours of sleep. And only had to switch rooms once. Still, when I met him out front the next morning he wasn't sounding too lucid. His sentences kept breaking up like ice floes, and his head, he said, was killing him.

"It's not as severe as the stabbing pain," he said. "And it's not as, like . . . mild as the vise-grip pressure. It's just . . . like . . . a combination . . . like a vise grip of nails."

By mutual agreement we pulled over at the Becker Butte lookout for our morning pill party. The stillness here was different than in Snowflake, as the land—cragged and turreted and bristling with mesquite and prickly pear—was much more turbulent. Far below, the Salt River petered through a deep gorge en route to the Roosevelt Reservoir. We were leaving the Colorado Plateau, falling off the Mogollon Rim into the lowlands around Tucson and Phoenix.

James popped the black case on the tailgate and began to rummage. Liothyronine. ("That's for my thyroid. A lot of us have thyroid issues.") 7-keto DHEA. ("It's supposed to help with fever blisters but helps other things, too.") Folic acid. ("Skin detox and brain supplement.") Montelukast sodium. ("That's just for asthma.") Cytozyme-PT/HPT. ("Has to do with the pituary gland.") Phosphatidylcholine. ("That's really good.") Amphetamine salts—empty. ("Adderall.") CoQ10. ("It's more for the brain. Wanna take one?") Copper. ("I'll take one of those.") Magnesium. ("Magnesium is really good.")

And countless others. Cholecalciferol, Kaprex Selective Kinase Modulator, SpectraZyme Metagest, B350, Meta I-3-C, Valacyclovir, PB 8, manganese, E-400 selenium, zinc, prednisone, Alprazolam, green tea, AdvaClear, Mitocore, Enzyme Optimization and pH Balance, Barlean's Olive Leaf Complex, potassium iodide, Methionine-200, L-carnosine, melatonin, coenzyme B complex . . .

James anticipated my question.

"How do I know all this shit works? I don't. I don't even know what all this shit is. I know when doctors tell me or whatever but . . . I don't . . . keep track of this too much. I have a NutrEval, like a blood test that shows the vitamins and nutrients that . . . I'm okay and that I'm deficient in. So, like, we adjust it based on that. Like, zinc is probably one of the most important . . . There's only, like, a few things . . . that I can really say I would, like . . . just grab and go that really work, that I can tell."

In the end James gulped somewhere between fifteen and twenty pills. If he wasn't detoxing he'd probably take fewer, he said. Getting into the spirit of things, I gobbled some modafinil and threw down some phosphatidylcholine and CoQ10 for good measure. Tinkering with consciousness, flicking it, probing it like a bruise.

It was reassuring to hear James admit that he couldn't detect the effects of most of the pills: it made his attunement more credible. It's not in other words like he was entirely lacking in a bullshit detector. For instance, he never bothered buying Extra Strength 5-hours because he couldn't tell the difference.

A lot of the more dubious pills had been sold to him a few months back by an alternative medicine doctor who claimed to know what supplements he needed by studying his genetic profile.

"I'm not saying she was trying to make money off me"—he said, hesitating, and then, finding no other outlet to the sentence—"but she was."

I perked up at this. It was the first time I'd heard him admit to anything like anger at anyone. He giggled uncomfortably but went on, the taboo having already been broken. "It's just so common. These so-called experts. They come up with a treatment plan and then also sell you everything that goes in the treatment plan."

As James's pique gathered, even Dave Asprey, the yak butter guy, came under fire. "He's another one who's got a company built around all this shit."

With James letting more and more emotion show it was beginning to feel like a breakthrough, and with growing interest I waited to see where it would lead. Then he took it in a direction I couldn't have anticipated.

"That's the fucked-up thing about this," he said. "It's always tied to money. So what if there just was no money anymore? I mean, we think we're advanced, but what if we were really advanced and there were some way we could all survive in the world without money . . . I used to say, well, money is necessary because it's, like, a civilized means of exchanging goods and services. But is it? I don't know."

It was a canny move. Money, like chemicals, made for a convenient villain. Both were everywhere, and both could be demonized without holding any one person responsible. It wasn't the doctor's fault. It wasn't the quack's fault. It wasn't Perkin's fault or Haber's fault. It wasn't anyone's fault. It was the money.

—

With about fifteen miles of gas left we coasted into Globe. As James pulled on his powder-blue latex gloves I went in to stock up on 5-hours, pausing only briefly to eye the magazine rack, which offered seven different gun magazines, including one dedicated entirely to the AR-15. When I returned, James was popping a Claritin. Yes, the pollen count was low, but his head was still killing him and Dr. Bernhoft, the L.A. doctor, had said that Tylenol and Advil would interfere with his treatment.

Heading southwest toward the Coolidge Dam and the crater-sized open pit mine near Safford, we pushed back out into the desert, the ridgeline of the Hayes Mountains like torn paper against the southern empyrean. The road stretched out like black taffy under the hot sun, paced by tumbleweed and old-fashioned telephone poles burdened by a single wire.

"This really is Looney Toons country," James observed, referring, I suppose, to the Road Runner cartoons.

The plan was to make for Tucson and then loop north. This would give Brian one more day to resurface before we went looking for him.

I'd been thinking a lot about Brian. In my head I suppose I'd always framed the journey as something like a long-distance doctor's visit, as if upon arrival I'd have some remarkable insight to offer, or be able to discern something about the true nature of EI that had escaped everyone else. But ever since my rambling arrival speech to Liz and Scott in Snowflake I'd begun to wonder if it was I who needed the insight, not Brian. And the hope I guess was that he, the outcast, the anchorite, the transfigured pariah, would be able to offer it.

It was a very old-fashioned notion. Not to mention romantic. I had long since given up on the idea that any single person could save me. Much to the relief, no doubt, of various romantic partners. Adulthood would always be, on some level, a lonely enterprise, and such gifts as one received from others could be sweet and consoling but would never be transformational. So I had always found, anyway.

Yet the vision of the transfigured pariah had been resuscitated by my waning faith in all the traditional paternal authorities—Medicine and Science and Government. The slow failure of these institutions, the strength of which had always been their monolithic anonymity, had somehow evoked their inverse. The enlightened recluse.

That this vision was an indulgence was obvious. Yet it was one that I could still afford. And—who knows?—some part of it might turn out to be true.

In the meantime we had set up a meeting in Benson, about one hour east of Tucson, with Dr. Gray, Dr. Hope's mentor—and Scott's doctor—who had agreed to an impromptu consultation. As the companionable afterglow of our twilight sardine-fest receded in the rearview mirror I was feeling the return of a frustrated desire for some sort of straight answer. What did TBI have to do with Lyme? What did mold have to do with chemical toxicity? Genetics, immunology, neurology, allergy, nutrition, psychiatry, environment . . . How did all of these modalities relate? And why was all of this so hard to know?

The great hope, of course—as Scott had intimated—was that there existed some kind of medical Rosetta stone which, once discovered and deciphered, would make everything clear. A Grand Unified Theory of human health. For sensitives it would be the ultimate vindication. Too easily overlooked, however, is what it would mean for the rest of us.

Consider, for instance, the implications of such a discovery on the structure of knowledge, medical training, and expertise. If symptoms from every traditional medical specialty could all be traced to the same underlying dysfunction, those specialties would no longer be able to operate in blinkered isolation. An entirely new way of practicing medicine would need to be invented to put the Grand Unified Theory into practice.

It didn't seem likely. Then again, everything that has ever surprised us didn't seem likely either, at some point. It's not an easy lesson to bear in mind. For as soon as the unlikely thing happens—or maybe a magical day or two later—the memory of its unlikeliness disappears.

History is rich with examples. Over 150 years before the term "EI" was even uttered, even before synthetic colors began to bloom across Europe, a similar wave of invalids, fleeing an equally mysterious ailment, descended on the desert Southwest in exactly the same way. And at that time the cure that would one day be found couldn't possibly have been imagined.

They came in the thousands, settling in tent cities on the outskirts of towns, wandering the streets in their bathrobes. And with the completion of the transcontinental railroad in 1869, and a spur reaching New Mexico in 1880, they began arriving by the trainload, stacked on stretchers and shoved through the windows of Pullman cars. In Tucson the tents reached nearly to the foothills of the Catalina Mountains.

Little was understood about tuberculosis in the mid-nineteenth century, though at the time it was the leading cause of death in the U.S. Not known to be contagious, it was, like EI, usually attributed to environmental factors, like damp valleys or unwholesome living conditions in overcrowded slums. The best way to restore the body to a state of purity was therefore to transpose it to a place of purity. As early as 1820 it was already well established that recovering from disease required rising above "the fever line" where the miasma could not reach. Several decades later one doctor was ready to declare that the location of this line could be found at precisely six thousand feet. More or less the same elevation as Snowflake.

In this way, geography (even road trips, according to one doctor) became, like a drug, prescribable. By 1884, doctors had banded together to form the American Climatological Association to advise on the medical advantages of various locales. To this end, they, like James, began collecting climatological data—altitude, humidity, temperature, sunlight, soil dampness, ozone, pine emanations . . . The Southwest in particular, with its arid air and high elevations, was believed to be ideal for drying out the wet lungs of consumptives—which is how tuberculosis treatment became the desert Southwest's first industry—once, that is, those unwholesome Apache and Navajo had been tidied up (many survived only to later succumb—in vast numbers—to tuberculosis).

It was known as the "triad treatment": fresh air, good food, and absolute rest. And by the latter half of the nineteenth century immigration boards and local boards of health were actively marketing it to attract new residents. "Come West and Live!" the slogan went. And they did.

Of all those who migrated to the Southwest in the nineteenth century fully 20–25 percent came in search of better health. It's what brought a tubercular Atlanta dentist named Doc Holliday, whose final resting place we had passed an hour outside Aspen. It's what brought a Wichita laundress, whose death in 1874 was the fateful fork in the path for her errant kid, Billy.

In the absence of a cure, consumptives became hyper-attuned to their own bodies, and this hyper-attunement gave rise (as it did with sensitives) to a unique subculture. Eventually, sanatoriums began to appear like great listening chambers where the illness could be better attended. In New Mexico alone, over 170 were built, predating, in their novel emphasis on cleanliness and ventilation, the EI housing that was to follow over a hundred years later. Indeed, so many sanatoriums were built that eventually aggressive zoning ordinances were passed to limit their number.

Then, in 1882, the reality of the disease suddenly shifted, with Koch's discovery of the TB bacillus. It was Benjamin Rush's dream come true, a way to exclude everything but the body from the medical optic—better yet a specific part of it, the smaller the better. In theory, a similar discovery awaited EI. Then again, it was equally possible that EI more closely resembled another mysterious disease that came into vogue soon after the tuberculosis mystery was solved: neurasthenia.

The two diseases made for an interesting comparison. One eventually achieved proper disease-hood while the other was discredited, teased apart, and—like Charcot's hysteria before it—absorbed into other emerging diagnoses.

These days, EI was most often compared to neurasthenia. And there were in fact a number of similarities—the endless jumble of symptoms, first of all. In his defining work on neurasthenia, *American Nervousness*, published in 1881, George M. Beard provided a nonexhaustive list of eighty of them, among them insomnia; hopelessness; fear of contamination; special idiosyncrasies in regard to food, medicines, and external irritants; vague pains and flying neuralgias; desire for stimulants and narcotics; lack of decision in trifling matters; pain in the perineum; and ticklishness. Speaking in 1905, one obviously irritated doctor (closely mirroring contemporary complaints about EI) referred to the diagnosis as so "elaborated and broadened and abused that today it means almost anything and with equal truth almost nothing."

And like EI, neurasthenia flourished despite having no known etiology. Beard offered a theoretical explanation that recalled premodern disease models, except in place of humors or "physiological excitement" (Rush's favored term) he substituted "nerve-force," attributing neurasthenia to a critical lack of it. And like so many doctors who treated EI, a comparable contingent who treated neurasthenia (including Beard—and, later, Freud) had also at some point suffered from it. Absent etiology, neurasthenia, too, was a disease you had to experience to understand. Which is probably why neurasthenics, like sensitives, were frequently dismissed as "malingerers."

But the greatest similarity by far was the cultural component. Like EI, neurasthenia was believed to reflect a specific cultural condition. "Americanitis," it was sometimes called, just as EI was called the "Twentieth-Century Disease." The culprit, Beard said, was "modern civilization" and the radical changes the country was undergoing at the time, chief among them steam power, the periodical press, the telegraph, the sciences, and "the mental activity of women." None of these were bad in and of themselves, Beard allowed, but they did overtax the nerve-force, and so increase the likelihood of depletion. The same was true of so many modern innovations. The wristwatch, for example, provided a convenient way to keep track of time, but only at the cost of a newfangled anxiety about being late. It was a construction familiar to anyone who has ever rued the invention of smart phones or email. Or even bathroom scales.

Even so, some scholars still argued that neurasthenia did serve a purpose, giving the inhabitants of the time a way to start metabolizing early modernity, a way to reckon with a radically shifting threatscape and the novel challenges it posed. (As one of the major contributors to this changing threatscape, it made sense that Haber should suffer from it.) And a way to start thinking about mental health as a totally legit and desirable goal. In doing so, it preheated the notion of psychological internality, anticipating Freud, without whom it would not have been possible to disparagingly impute subconscious motives to victims of EI a hundred years on.

Even if Beard's model of human health was painfully simple, in other words (he literally compared the human body to a light bulb), and his diagnosis hopelessly broad, he was clearly onto *something*. Which begged the question: Even if it did turn out to be humbug, was it still possible that EI served some deeper cultural or ideological purpose? To help us reckon with a threatscape that was once again radically shifting? Or—like quackery, but more broadly—to help us reassert the human perspective that was always being shunted aside by wormy politicians, revolving-door regulators, and shortsighted corporations that some shithead judge decided were people?

In this view, disease emerged in the body like some kind of foreboding dream, a negative physical gestalt of all that was going wrong in the world around us. In the daylight hours it haunted the periphery of our consciousness, attuning us to the dangers that we were disinclined—and perhaps politically or culturally unable—to acknowledge. In society at large, this dream was felt as a tremor traveling the edge of our collective subconscious, like the cultural fixation with apocalypse movies. Could these tremors tell us something? Could they be deciphered?

Beard seemed to think so. And some EI researchers did, too. It was a variant of the canary-in-the-coal-mine argument—a collective shudder of somaticized dissent from 12.8 percent of the population. At issue, after all, was, in the words of one outraged EI researcher, "a culture that does 'risk assessment' of each of our chemicals [and allows] a certain number of people to get sick or die from exposure to each." Surely the calculus of a diseased body politic. Even the institutions that were supposed to protect us from contamination were "created out of the same industrial paradigm that allows this contamination," and were therefore "not only not positioned to respond in any constructive way, but . . . in many cases set up to deny and distort the reality of chemical-induced disability."

Civilization itself, in other words, was in the grip of a vast psychosis which not even the conscious mind could penetrate. Instead it was left to the subconscious mind—and the subconscious body, like the twitching limb of a disturbed dreamer. As with the "deployment-related stress" of Gulf War soldiers, it was the body's way of saying, *Something here is very, very wrong.*

Ultimately, though, while the neurasthenia comparison did grant EI a certain cultural legitimacy, it still denied it clinical legitimacy. Whether you accepted the comparison depended on where you were coming from—just as scientific knowledge depended on who owned it, and risk assessment depended on who did the assessing. Those with no sympathy for EI whatsoever were more likely to deny it both clinical and cultural legitimacy, and write it off as mass hysteria: a delusion induced by sensationalizing news outlets and sustained by social media.

The hysteria argument wasn't particularly hard to make. The suggestibility of the human mind was well established. The greater part of our beliefs, in fact, were influenced by what others believed to be true, as research showed. In one study I'd read, for instance, subjects were told that they would be testing a new antiviral pill designed to be used in the event of a flu pandemic. The purpose of the test was to ascertain whether the new antiviral (actually a placebo) had any side effects. Among the experiment's subjects were two actors who feigned headache, abdominal pain, and other symptoms after taking the pill. Later analysis showed that subjects exposed to the actors reported eleven times greater symptoms than controls.

Not that this indicated some kind of personal failing or character flaw. On the contrary, taking cues from others was usually adaptive, a time-saving heuristic to help us get our bearings by leveraging access to a dispersed network. But misreadings did occur, and while it might not be possible to say whether EI was one of them, it was certainly possible to identify the factors that made it more likely. One was the availability of reliable information. Whether the answer was actually known. Cheating, for instance, was a fine test-taking strategy, but only if someone in the room knew the answers. But when it came to EI no answers were known.

This led to the second factor: the prevalence of misinformation—or hype. Take the gluten craze. Strictly speaking, gluten was only dangerous to the 1 percent of the population that suffered from celiac disease. Yet surveys showed that 30 percent of the population was trying to avoid it. Doubtless this would not have been the case if the issue hadn't received so much media coverage.

The final factor was emotion—fear, primarily. In the 1950s, for instance, in a little town about an hour and a half north of Seattle, tiny pits began appearing in car windshields. The news media picked up the story, feeding fears that the pits were somehow connected to atomic fallout from hydrogen bomb tests being conducted in the Pacific Ocean. Within days, police all over the country were receiving reports of pitted windshields. In the end the pits were recognized as typical wear and tear from road debris. But given the nuclear threat you could see how one's attention might be drawn to the flimsiness of the barriers we trust to protect us. Because then at least you could do something about it. Then at least you could call the cops.

A more recent example was chemtrails, those spumes of evil secretly administered by the government from the rear end of high-flying jets. Why? Well, to test experimental new compounds on a human population, of course. To scramble our fertility. To weed out the old and infirm.

Yet "mass hysteria," while accurately capturing a very real media phenomenon, did less well accounting for the physical symptoms of EI. There was a difference between brain fog or an icepick headache and the belief that atomic fallout was pitting your windshield. "Mass psychogenic illness," a related term, more ably addressed the physical symptoms but, unlike EI, was mostly limited to small, cohesive groups (girls' private schools, commonly) and furthermore tended to be relatively transient. Those afflicted eventually snapped out of it. Sensitives were not known to snap out of it. On the contrary, something about EI pulled them in ever deeper.

It remained possible that EI was neither a new kind of neurasthenia nor mass hysteria. That it was, like tuberculosis before Koch, a legit disease cunningly disguised and waiting for someone to reveal it. After all, dismissing EI because it couldn't be explained implied that it was only possible to suffer from that which was already understood. In which case there was a great deal to be said for ignorance.

It reminded me of a story I'd heard about a woman convinced she had suddenly become allergic to meat. It happened one day after grilling a leg of lamb with her neighbors. A few hours later her stomach began cramping and she began feeling light-headed and anxious. Not thinking much of it, she splashed water on her face and figured she had maybe accidentally eaten a snail adhering to the wild onions she had grilled with the lamb. Two weeks later she was eating a cheeseburger when she again began feeling weird. Then she noticed hives appearing all over her hands and stomach. Making the connection, she concluded she'd somehow developed an allergy to meat. But a doctor told her in no uncertain terms that "adults do not just suddenly become allergic to something they've eaten for forty years. And certainly not meat." In other words, it's all in your head.

A Virginia-based allergy specialist named Thomas Platts-Mills had been telling people the same sort of thing for twenty years. And he would probably still be telling them the same thing today if it weren't for another mysterious allergic reaction. The difference is that this one involved a cancer drug backed by a deep-pocketed pharmaceutical company. Teaming up with the company, Platts-Mills soon traced the allergic reaction to a particular sugar attached to the drug. Which still didn't explain why some patients reacted and others did not. In mapping the cases, Platts-Mills soon discovered that they were limited to certain parts of the country. It seemed like the beginning of a pattern, what might have been a figure in the static. Handing the map to a technician, he told him to look for any map that might match. And eventually the technician found one: an incidence map for Rocky Mountain spotted fever. A tick-borne disease.

Still, it wasn't until Platts-Mills stumbled into a nest of ticks himself—and then, after feasting on lamb chops, woke up six hours later feeling light-headed and anxious—that he made the connection. Red meat contained the same sugar as the cancer drug. And the reaction to it was primed by a tick bite.

The story, a classic of its kind, illustrates several points. First, the patient reported the illness first. The patient *always* reports the illness first. So the report should never be discredited for that reason. Second, the discovery was not the product of methodical research but (like Perkin's discovery of Tyrian purple) purely a matter of chance. Meaning any scientific culture that is inhospitable to chance is bound to be less fruitful. (Only a fool expects new knowledge to be unsurprising.) Third, it wouldn't have happened if the researcher hadn't experienced the disease himself, in effect becoming a patient. Meaning experience can be more enlightening than knowledge, and knowledge (expertise) has a tendency to get in the way of experience, and even to regard itself as adversarial to experience, which it dismisses as subjective. Fourth, it wouldn't have happened if the original prompt—the cancer drug—wasn't already considered a legitimate, financially viable object of scientific study. Any self-respecting scientist, in other words, would do well to remember that (echoing Rush's defense of old wives' tales) the truth is as likely to emerge from "illegitimate" objects of study as legitimate ones.

Dr. Gray had not experienced EI personally. Yet he did have the next best thing, which was decades of clinical experience in environmental medicine—plus early training courtesy of William Rea, environmental medicine's most prominent practitioner. He also had a solid background in mainstream medicine, and worked as a hospitalist at a local hospital. In other words he seemed to have credibility in both worlds. If anyone could distinguish whether EI was more neurasthenia or tuberculosis, it would be he.

CHAPTER 17

The Gospel of Dr. Gray

W e blazed south into the flatlands, the blue range of the Pina-
leño Mountains to the west and to the east just thirst. On the
outskirts of nowhere a sign raced up to say "Thank You" for
no obvious reason, except to make us realize how long we'd been waiting
for this simple courtesy. Then, disappointingly, a few hundred yards farther on, another sign: "For Visiting Willcox." We had, evidently, just visited
Willcox.

Onward, past a dry lake bombing range which didn't show up on any
tourist map, through the Sulphur Springs Valley, riding the hot meridian
down toward Nogales. James had put on another episode of *Philosophize
This!*, and our host was chattily unpacking Voltaire as we descended into a
valley strewn with truck-sized boulders. New variations of cactus were beginning to appear, thin and antlery, pineapple squat. "Doubt is an unpleasant condition," our host said, quoting his subject, "but certainty is absurd."

The road swung north and we exited onto a wide empty street, past
tattered trailers and self-storage warehouses. A slow right at the deserted
86 Cafe and then another, down an even emptier street, past dirt yards

and shoebox houses and trailers like desert freighters riding out the sun. The Rover slowed to a drift, James's talk of buying and renovating his own trailer park fading as we began to wonder whether the GPS had led us astray.

A call to Dr. Gray confirmed that it had not. Following his instructions, we proceeded to the end of the road and hooked a left down a long driveway. A low yellow house waited at the end of it, with an ancient bus parked on the grass, the bottom half painted a teal that might have once been green. To the right, a tilled field baked in the heat, at the edge of it an iridescent blue Volkswagen Beetle with dune buggy wheels and a large overturned flower pot in the front trunk. One part of the yard was surrounded by a high chain link fence, and beyond that we could see what looked like a shipping container. None of this made any sense. We parked under a tree and cracked the doors on throbbing brightness and 100-degree heat.

———

"This is Jack," Dr. Gray said.

Jack was four feet long, forty years old, and living in the fenced area of Dr. Gray's backyard. He was a tortoise, an African sulcata, actually, native to the southern Sahara, and in tortoise years hadn't even reached middle age.

"I'm going to have to provide for him in my will," Dr. Gray said.

Jack originally belonged to one of Dr. Gray's patients. The patient also owned a duck who liked to ride on Jack's back. And all was well. Then one day Dr. Gray's patient asked him if he would be willing to look after Jack while he, the patient, did some traveling. Not thinking much of it, Dr. Gray agreed, because that's the kind of doctor he was.

The real story was that the patient had grown weary of the alarm clock jolts from his defibrillator and had decided to switch it off and wait for the last lights out. Two months later it arrived at the local Walmart, where he collapsed and died. That was three years ago. And Dr. Gray had been feeding Jack a lettuce a day ever since.

What happened to the duck is not known.

Even without Jack, it would have been obvious that Dr. Gray was no ordinary doctor. About five-foot-five, sixty-five years old, and of clear Native American extraction, he wore sandals and white knee socks. His knees had a view, but then came khaki shorts pulled up to a cute little paunch, and a blue-and-white checkered shirt over which he wore a khaki fisherman's vest. A bush hat bedecked with a peacock feather and a Bernie Sanders pin topped the ensemble, with a little white ponytail protruding from the rear.

Dr. Gray appeared to be one of those small people with access to huge stores of energy. His talk went a mile a minute, all of it highly erudite and remarkably articulate, but possessed of an engrossed quality such that efforts to redirect it felt not unlike trying to get the attention of a jackhammer operator. Here was a man who had absolute, unimpeded access to every square inch of his considerable knowledge base, but perhaps less of a handle on how to package or summarize that knowledge base—or else, being legitimately fascinated by his subject, he simply didn't want to.

We had not known that we would be meeting Dr. Gray at his private residence. But then, we had not known he owned a four-foot tortoise, or a one-and-a-half-acre vineyard, or another twelve acres of alfalfa. Dr. Gray was an unusual man. He'd been in Benson since 1986, initially as a general practitioner boarded in general preventive medicine and occupational medicine. But after several years of offering free services to anyone unable to pay, he donated his practice to start a nonprofit clinic which could more easily obtain grants to serve the indigent. Today he worked part-time at the clinic, seeing toxicity cases once a week. He had also co-founded a nonprofit ambulance service. All of this boded well by suggesting he was a man who hadn't allowed his expertise to occlude those it was meant to serve.

After introducing us to Jack Dr. Gray took us back into the kitchen, offering a choice between apple cider, 70-calorie organic limeade with mint, or a fresh pot of coffee. It was about 30 degrees cooler inside. Hardwood cabinets lined the walls interspersed with jarred grains on open shelves. A fridge magnet held a faded picture of a tortoise and a duck. Dr. Gray, the darkness fading from his auto tint glasses, seated himself in a well-cushioned wheeled wicker chair and rolled up to a laptop on the kitchen table. He knew what we had come for and promised he could deliver: a unifying hypothesis.

"So this is your own theory?" I said.

He hesitated.

"I definitely subscribe to the science that I'm going to describe."

Hmmm.

"I don't want to take ownership of the sum of what hundreds of thousands of people have done," he explained. "I just want to distill and organize it for you in a way that fits with my clinical experience. And I will tell you that if you had asked me the same question thirty years ago I wouldn't have been able to answer."

He gave a deep sigh, and began.

"We have to go back and look at the development of life on the planet."

I, too, gave a deep sigh, and began to jones for a gulp of 5-hour.

It was a 2.5-billion-year project, Dr. Gray said, and it never would have moved past year one if those early life-forms hadn't learned life's first rule of thumb: Don't eat yourself; eat others.

He was talking about the immune system. The means by which the body distinguishes between self and not-self.

"So that's one piece. The other piece is that even in the days of one-celled organisms one species after another has thrown poison darts at each other, and those darts, if effective, would wipe out other species. The only species that survived were those that had somehow developed an enzymatic mechanism to neutralize the toxin. The code for that enzymatic mechanism is stored linearly in the DNA and gets passed from generation to generation. If we audited every advanced organism living on the planet today—plants, animals, probably most of the fungi—we'd find that they have the ability to make eighteen thousand different cytochrome p450 enzymes, which are the repository of the detoxification array. And that really represents the pool of most of the natural poisons that exist on the planet."

"So it's a genetic disease?" I said.

"Well, it's a genetically based illness," Gray said. "I agree with that. It's a genetically based illness in a social environment in which human beings have made a lot of bad decisions."

The problem, Gray said, was that human activity had enlarged the poison pool by orders of magnitude. Plus, we'd amplified the mutagenic load by orders of magnitude—meaning we'd increased the amount of stuff (mostly chemicals) that threatened the stability of our DNA, and thus our ability to manufacture the enzymes necessary to eliminate toxins—our "detox array." According to Ritchie Shoemaker, a legendary figure in the mold community, an iconoclastic guru with, by all report, highly limited interpersonal skills, somewhere in the neighborhood of 24–28 percent of the population was missing significant elements of the detox array. The elements they were missing determined which toxins they were unable to clear. And the toxins they were unable to clear determined which systems were impacted. Organophosphates, for instance, impaired the autonomic nervous system. Toxic mold impaired protein synthesis and mitochondrial function. Estrogenics like dioxin and BPA impaired the endocrine system. All of them could fire up the immune system—and fired up is how it remained, since the triggering irritant couldn't be cleared. And, depending on which systems were impacted, the symptoms would present as a different disease.

"Multiple Chemical Sensitivity is a subset," Gray said. "It is actually a massive problem, reaching to Parkinson's, dementia, seizure disorders, autism, ADD, multiple sclerosis."

It was the same sort of thing that Pall said, the same thing Miller said: not just one disease but an entire family. A claim which not only overturned the current nosology but significantly reoriented disease around external factors like toxins and mutagens and presumed a solid genetic basis. The grandiosity of this vision rang all kinds of alarm bells.

True, tentative research findings had been made linking EI with eighteen different genes, including the MTHFR gene, which had also been linked to migraines. But as with every other EI biomarker these findings were frustratingly inconsistent, and always bordering on clinical insignificance. Marginally more convincing genetics work had been done with myalgic encephalomyelitis, but the case was nowhere near clear-cut. If mapping three billion base pairs of the human genome had taught us anything, it was that the causal relationship between health and genetics was complex.

By this point we had moved to the living room, which with its spacious layout and cathedral ceiling gave the impression of being somehow, magically, larger on the inside than the outside. Rust-colored couches formed an L around a tree stump coffee table. Stained glass frames filtered the light from windows on either side of a stone fireplace. A large harp stood in one corner and a piano in another, but the only tune came from a wall-mounted dehumidifier that kept up a dull hum.

Gray was explaining how sequestrants worked, latching on to toxins and flagging them for removal from the body. The problem was that sequestrants (like clay, glutathione, and charcoal) could only capture toxins at large in the body—not the stuff that had accumulated in the fat, bone, and connective tissue over the course of a lifetime. For this reason, it could take two to five years on a steady diet of sequestrants to completely clear the body of toxins. And that's if you were mono-susceptible—missing only one piece of the detox array. For the multi-susceptibles the news was much worse.

"The multi-susceptible individuals can't stop detoxing unless they go to the South Seas and isolate themselves from all exposures. Because the exposures are ongoing, and they will always get bombarded."

Here James cut in, saying that he'd detoxed down to zero several times.

"But I'm still hypersensitive," he said.

"You'll remain hyper-reactive the rest of your life," Dr. Gray pronounced without hesitation. "That is not going to change, I'm sorry about that."

James groaned and tipped his head forward into his hands. Seeing him so wounded made me want to push back, so I asked Gray why his views weren't shared by the rest of the medical world. His answer was simply that no one had told them. He cited a 1980 survey in which every dean of curriculum in every medical school in the country was asked two questions: Do you have required time in your curriculum devoted to occupational medical health; and, if yes, how much?

"Seventy percent said no," Gray said. "And of the 30 percent that said yes, the average clock-hour time was 4.2 hours. Over four years. I trained at the University of London in the Institute of Hygiene and Tropical Medicine, and the average clock-hour time on this subject in Europe is sixty to eighty hours. So the subjects are just not taught."

Not only that, he said, the survey was redone ten years later and it was up to 6.2 hours. But sixteen years later it had dropped back to 4.2. I asked why this was.

"The emphasis right now in research is pharmaceuticals," he said. "If you don't get money, you don't do your research. The primary funders of medical research in this country right now are the drug trials of the pharmaceutical companies. I don't know what all of the other factors are but I can tell you this: in Europe, 60 to 80 percent of the workforce are in trade unions, and they have a voice. And they're involved in the political and social processes, and the distribution of resources. That doesn't happen here."

I was still trying to get a handle on Gray, and figure out how to take everything he was telling me. So far, he sounded highly credible. He was erudite, he acknowledged when he didn't know something. He wasn't selling any kind of product, and in fact had made sacrifices to help the indigent. Plus, for two years he had been taking care of a dead guy's tortoise.

The only marks against him were his tendency to lose track of his interlocutor as he chased his subject matter down increasingly obscure corridors, and his rather grandiose belief in the role toxins played in modern illness. But his fixation on his subject matter wasn't necessarily a bad thing. And as for grandiosity, he had been very clear up front that these ideas weren't his, and in fact had gone out of his way to give most of the credit to Shoemaker.

I decided to add a bit more heat, acknowledging the role played by medical schools but observing that this still didn't quite amount to active resistance. Gray was undeterred, offering by way of reply the story of why he left academic medicine twenty-seven years ago—his origin story.

The story began in 1973, in Globe, Arizona, where we'd been stocking up on 5-hours only a few hours before. In 1973 Globe had been home to three asbestos mills. Two of them were shut down by local regulators for failing to meet EPA dust emissions standards. One of the mills went dark immediately. The other was owned by a guy named Jack Neal, who kept it open for another year, a renegade operation that ran only at night. Regulators eventually caught up with Neal, but in the meantime he had somehow convinced the Globe City Council to rezone the property for residential development—despite an explicit warning from the county regulator that the unremediated piles of asbestos on the site could cause illness and death. Neal, who, real-estate-developer-wise was basically the opposite of James, solved the problem by bulldozing the asbestos and covering it with ten inches of topsoil. Then he subdivided the property into a fifty-five-unit trailer park he dubbed "Mountain View Estates."

By October 1979 the sewage at Mountain View Estates was backing up into the units because Neal had cut corners and only installed one pump. A state sanitation inspector sent to investigate soon discovered that the entire subdivision was sitting on a layer of asbestos the size of two football fields. The inspector duly reported his findings to his superiors at the Arizona State Health Department, who responded with a letter to the residents. Gray described the contents of the letter as follows:

"This is to advise you that there may be asbestos contamination in your subdivision. We are investigating. It's going to take us about three months to complete this investigation and during that time we recommend that your children not play outside and don't wear your shoes in the house. Okay?"

The letter, Gray said, was handed to each family by a Health Department lackey wearing a full face respirator.

At this point, two months after the asbestos had been discovered, Gray was contacted by the National Institute for Occupational Safety and Health (NIOSH), a division of the Centers for Disease Control, who asked him to investigate. Gray, who at the time was working at the Arizona Division of Occupational Safety and Health (ADOSH) at the University of Arizona, quickly issued a report recommending that the 130 residents be evacuated immediately. The report leaked to the press, and Gray (as I later discovered on researching the incident) was quoted in *The Arizona*

Republic calling the State Health Department's actions "criminal" and saying that they should have evacuated the community the moment the asbestos was discovered. The same newspaper story also cited the director of the environmental services division of the Health Department as saying that he would "consider" Gray's comments.

The day the story appeared Gray was summoned by the director of ADOSH and fired. One month later the governor of Arizona declared a state of emergency in Globe, but after two years asbestos remained on the site and no one had been evacuated.

"We never promised anyone money to relocate," an assistant to the governor was quoted as saying.

Gray was eventually reinstated at ADOSH thanks to an intervention by NIOSH, who requested that he lead a more thorough investigation. But he only lasted another year and a half in the academic setting.

"After that I didn't have a place to see patients," he said. "So there was a concerted effort. I don't know who was doing what. All I know is they squeezed me out."

It was a credible enough story. A case study in all the ways that greed and power impede not only justice but common sense. The ending sounded a bit vague, however. Was Gray the upstanding whistle-blower he presented himself to be? Or was he a nettlesome, self-absorbed malcontent masquerading as a victim of "the powers that be"?

The point was kind of moot with the lives of 130 people at stake, but I trolled him anyway, observing that it was pretty easy to drift into conspiracy theory.

"I agree with you," Gray said. "So I don't have a theory. But what I do know is this. I moved over to the county hospital. I set up an occupational medicine program there. I ran it for the county, and I was the Pima County physician for six or seven years. But I constantly was being interfered with . . . So I decided to come down to Benson and set up my practice in a community that was underserved. Because I knew that if I was there for my community in the middle of the night when they weren't insured they wouldn't care that I was doing controversial research or getting involved in complicated court cases."

You're one of the good guys, I said.

"Well," Gray said, "I try."

We collected ourselves and were heading for the door when Gray arrested us with a final warning. According to EPA data published two years ago, the area between Tucson and Phoenix, where we were heading, had the dirtiest air in the country.

"One fifth of the nation's cotton is grown between Tucson and Phoenix," Gray said. "Sixty-five percent of all pesticides in the country go on the cotton crop. Roughly 12 to 15 percent of the entire burden of pesticides used in the nation is going onto that soil—and then being churned every two to three months with every growth cycle. And then the dust storms come. But even without the dust storms, the dust particles that become aerosolized carry chemicals and poison."

Nor could we just close all the windows and put the vent on recirc.

"Even when you close the vent," Gray said, "you're still getting 10 percent external air by law."

CHAPTER 18

Prescribing the North Wind

We walked out into blazing heat under a crisscross pattern of fading chemtrails. The high sun made dark caves beneath the trees and from the south came the thunder of a storm rolling up from Mexico carrying with it God knows what.

It was mid-afternoon and we still had another seven hours of driving to reach Brian's location. The alternative was to shoot for Sedona, a four-hour drive, and try to find Brian tomorrow. This spared us a midnight arrival in the middle of nowhere, and gave us enough breathing room to stop and see David Reeves, the publishing guy who had compared EI to pregnancy.

Reeves was a patient of Dr. Gray's, but, more importantly, of the dozens of sensitives I'd spoken to, he was the one with whom I identified the most. He was, therefore—to me, anyway—the most credible, and I was eager to see if anything about him seemed off or rang false. It was like the impulse you get to look in a mirror after being wounded somehow. Would I recognize the figure staring back?

J sat silent behind the wheel, resquaring himself against fate. He seemed to have mostly recovered from the news that he would never be free of EI, but when he did finally speak, his voice had dropped to a barely audible whisper.

"If I'm always going to have sensitivity that's really depressing," he was saying. "But . . . we should follow up with Bernhoft. According to Bernhoft, with consistency . . . you can regain . . . his patients have regained their health. Bernhoft treated himself this way . . ."

It was like the voice of a child feebly clinging to some cruelly shattered hope. We drove on in silence. Feeling myself beginning to fade, I popped a tepid 5-hour and gnawed a lamb bar. We'd asked Gray what it was about lamb that made it so uniquely tolerable to sensitives. At first he didn't have an answer, but then he remembered a study that had been sent to him by a veterinary medicine doctor in the Department of Agriculture. In the study, one thousand heifers were fed grass and one thousand were fed Roundup Ready soy. They all became pregnant. The grass-fed heifers had zero cases of fetal demise. The soy-fed heifers: 449.

"So they're getting a lot of glyphosate in the soy," Gray said. "I don't know whether that has any role to play or not, but I do know that your sheep are going to be grass-fed."

Like most things, it was possible. But by then I had become accustomed to the presence of these lesser unknowns. There was even about them something almost fondly familiar, like a chipmunk who lives beneath the porch and can every now and then be drawn forth with an almond. I was beginning to think that it had never been in the nature of truth to reveal itself. The real question was, where did we ever get the idea that this should be otherwise? That the truth should oblige us and sit there clearly labeled, winsomely awaiting inspection like any other packaged product on the shelf?

True, there were certain ideal, starlit conditions in which truth could sometimes be found. But these conditions were a lot rarer than most people liked to believe. This, I think, was the point of Gray's story about why he quit the academy and built his life's work in a place where power and money were, like air pollution, absent. Because it was only in places like this that the truth could be found. Everywhere else it stayed hidden. By state health departments. By town councils. By greedy developers, bureaucracies, governors. By whoever had an interest.

And this led to another thought. I'd been accustomed to regarding guys like Gray much as they were regarded by the establishment: as medical crackpots working out of their garages using unorthodox methods to arrive at unreliable results inflated to compensate for their lack of institutional standing. In other words, kooks. But in reviewing all the figures that Gray had credited with helping him arrive at his current understanding— specialists like Bill Rea, family physicians like Shoemaker, psychophysiologists, industrial hygienists, occupational medicine physicians, infectious disease physicians, a guy who operated on hernias—it dawned on me that what they all had in common was that they were all, first and foremost, clinicians. In aggregate they comprised something like an underground network of unaffiliated researchers whose mode of truth-seeking differed radically from that of their academic counterparts. While academic researchers spent their careers conducting narrow studies on usually somewhere between thirty and two hundred subjects, for instance, the clinicians spent their careers developing a much broader understanding based on a cohort numbering in the tens of thousands: their own patients.

The divide between the two camps was, like the divide between mind and body, so profound that it hardly needed to be mentioned. You simply knew which side you were on. What's more, each side seemed to have an equally low opinion of the other. The academic researchers regarded the clinicians as amateurs, lacking in rigor as well as a solid grasp of the scientific literature. The clinicians considered the academics to be irredeemably biased by professional ambition and pharmaceutical funding sources. In their view, human health was a human issue, and was therefore best pursued holistically, in the context of actual humans.

Shoemaker had once defended himself along these lines, arguing that, compared to academia, which only considered "one small facet" of the problem, the "systems approach" of the humble family practice actually offered the better paradigm for studying "the immensely complex world of ecological health."

And maybe this is what irritated Gray's employers so much that they had to fire him: he put the patient first—ahead of the protocol, ahead of the bureaucracy, ahead of the politics, and, worst of all, ahead of his own self-interest. After all, if you cannot rely on someone to pursue their self-interest it becomes a lot harder to control them. You might even say that this was one of the kook's defining characteristics: the refusal to be controlled. And whatever is uncontrollable is also a threat.

The "system" by its very nature could not be a kook. It could, however, be corrupt. Thus in the absence of certainty the old question returned: Which was more likely? That the system was corrupt? Or that Gray was a kook?

———

You couldn't ask for a loopier—or more uniquely American—illustration of the complexity of this question than the case of William Rea, the éminence grise of environmental medicine. Rea began his career as a thoracic and cardiovascular surgeon but became interested in clinical ecology in 1970 when his son developed severe asthma, and all the standard treatments failed to offer either explanation or relief. When he learned about Theron Randolph's work in Chicago, he reached out to him and began using his protocols to treat his son. Around the same time, Rea began developing symptoms of his own and was soon living out of a tent in his backyard.

After meeting with Randolph in 1974, Rea opened his own clinic, the Environmental Health Center, in Dallas. His treatments followed an allergy model in that he tested for allergens and then prescribed ongoing injections of a very low dose of the allergens for which the patient tested positive (an antigen). The immediate goal was to switch off the allergic reaction, with the longer-term goal of hopefully desensitizing the immune system altogether. Except Rea did not limit himself to recognized allergens. The list of substances for which he tested was 259 items long and included everything from men's cologne to rug padding.

"Rea helped us appreciate the reactivity of the immune system beyond allergy," Gray had said.

Rea's practice grew over the years. By the time I learned about him he claimed to have treated over thirty thousand patients, including factory workers, pesticide sprayers, Gulf War vets, and 9/11 first responders. Rea's reputation grew as well, particularly in Europe, where, like Hahnemann before him, his views were embraced by the aristocracy, among them the Duchess of York and Princess Anne. Rea went on to author multiple textbooks on Environmental Illness and serve as a mentor for hundreds of doctors—including Gray. When Randolph died in 1995, Rea naturally took up his mantle, and by 2007 Gerald Natzke, the president of the American Academy of Environmental Medicine, which Randolph had founded over forty years earlier, was calling Rea "a legend within our ranks."

But Rea's prominence attracted an equal measure of enmity. Mainstream doctors accused him of being a quack and fleecing his patients. I'd even talked to an EI doctor who, after describing Rea as "the king" who "saved us all," admitted that patients often spent $30,000 at Rea's clinic and still didn't get well. As proof of his fraudulence, Rea's detractors often cited the fact that he had been disciplined by the Texas Medical Board. And in fact he had. In 2008, ABC's *Nightline* covered the story in a ten-minute segment called "Dangerous Doctors."

Rea defended himself in a letter to his patients claiming that he was the victim of a vast conspiracy (for what other kind of conspiracy was there?). "To put it bluntly," Rea wrote, "there is currently an organized nation-wide effort to destroy the specialty of Environmental Medicine and to eliminate from practice physicians who diagnose and treat patients suffering from chemical sensitivities. . . . This campaign has been going on for at least 10 years now and is being led primarily by health insurance companies."

Maybe this was true, and maybe it wasn't. Certainly it could have been. As Gray had said, "The insurance lobbies have a way of turning the old-boys network against people who are threatening the status quo."

But unlike Gray's speculations regarding the tolerability of lamb, which I was content to leave unconfirmed, peacefully grazing on the mountain of unknowability, in this case it seemed reasonable to think that the truth could be known. And probably should be, too.

At first glance, the charges leveled against Rea by the Texas Medical Board (TMB) did indeed seem pretty dire. The tests he ran on his patients did "not even qualify as experimental, and are more properly described as pseudoscience." The SPECT scan reports, for instance, "all read word-for-word the same for all patients"—not a good sign. What's more, the diagnoses Rea gave were sometimes entirely imaginary, like a syndrome he called "vasculitis-radix paradox," which, though pretty cool sounding, had zero basis in the medical literature. And his treatments—like the injection of jet fuel—were "potentially dangerous." In fact, given his lack of formal training in allergies, toxicology, or genetics, it was doubtful that Rea was even qualified to practice. The only board-recognized training he had had was in thoracic surgery.

Worse still emerged upon perusing Rea's 292-page deposition. In it, he clarified that he did not inject his patients with jet fuel, but rather jet fuel exhaust, which was collected behind the chain link fence at the local airfield. By way of offering further reassurance, he explained that, actually, by the time he had finished diluting it, the antigen he used contained no chemicals whatsoever—only their "electromagnetic imprint." Which is where things began to get a bit silly. Rea then went on to admit, in a wonderfully straight-faced way, that he sometimes had occasion to treat people with an antigen derived from "the north wind."

Now, really reaching, I suppose you could imagine a scenario in which someone lived downwind of a chemical factory, and collecting an exudation from said factory might be necessary to test that person's reactivity. But further questioning revealed that Rea did no such thing. In fact, the north wind he collected came from right outside his clinic. Which seemed to suggest that Rea seriously believed that there was something about the "northness" of the wind—quite apart from what it might be carrying—that was capable of impairing human health.

On reading this I admit I had mixed feelings. The first was a giddy impulse to stand up and applaud the sheer lunacy of it. There was after all a certain poignant grandeur to it, like the slow-drifting blimp that made that woman's toe twitch. Or Hahnemann bent over his eyedropper, dutifully diluting some herbal decoction out of existence. By all means! Let us collect the north wind, let us summon the south as well for an impromptu consult in the parking lot! Let us derive the essence of passing cars, and pitch nose hairs over our left shoulder! Why not, when the alternative was a dour, profit-driven medical bureaucracy that, though dedicated to saving life, knew nothing whatsoever about what life was actually about? The only thing missing now was music and dancing, firelight and face paint. Let us return to these things, lay hands on each other, and call out to the spirit! Call out to the north wind!

Eventually, reason returns with its thudding step. We master ourselves and sigh. For the truth was that Rea's prescriptions were little more than a mirage of placebos distinguishable from the straight-up superstitions of sensitives only in their pretense of medical legitimacy. Not that there was anything wrong with placebos, per se. But Rea seriously seemed to believe that these were not placebos but real, pharmaceutically efficacious treatments. Clearly, this was a case of kookdom—not corruption.

Yet digging a mere inch or two deeper than *Nightline*'s shallow reporting revealed that it was not so simple. For instance, several days before Rea's letter to his patients it came to light that the chairman of the Texas Medical Board's disciplinary committee, a man named Keith Miller, was at least as compromised as Rea. Not only had Miller testified against doctors in over forty medical malpractice cases while on the TMB, he was also getting paid by Blue Cross/Blue Shield to sit on their advisory committee—a clear conflict of interest. These revelations were followed by Miller's immediate resignation, and then further allegations from the American Association of Physicians and Surgeons (AAPS) that Miller was in fact a straight-up stooge of the insurance companies, having used his position on the TMB to sanction practitioners of environmental medicine (including Rea) at their behest. And that the five patients on which the complaint against Rea had been based (1) didn't even know that their cases were being used for this purpose; (2) actually testified in *defense* of Rea at the TMB hearing; and (3), most damningly, all just happened to carry the same insurance—meaning that the complaint against Rea in all likelihood originated with an insurance company, meaning that he was indeed being purposefully targeted, just as he had claimed. The AAPS suit also contended that the board had ignored the testimony of these five patients, along with the testimony of seventeen experts in environmental medicine, in favor of a single anonymous expert who wasn't even in Rea's field.

With that the diagnosis seemed complete: corruption—not kookdom. And there was more to come. For the furor that followed the Miller revelations begat an epic eleven-and-a-half-hour hearing before the Regulatory Subcommittee of the Texas House Appropriations Committee. The original purpose of the hearing was, as the chairman put it, to review the "medical board's operations from the fiscal perspective." But comments made by the chairman early on ("We just got a call, folks, from one of the golf courses wondering where all of their doctors were today.") suggested that the room was packed (in fact an overflow room had to be prepared) and that some irregularities might be in store. Even so, no one could possibly have anticipated that these irregularities would include prominent members of the Texas Medical Board being publicly excoriated by legislators and the board itself being twice compared to the Third Reich.

The first two hours of the hearing passed unremarkably, mostly living up to what the chairman had proposed. Everything changed, however, when the first member of the public arose to give testimony. Her name was Sharon Marie Fuentes and she had been endeavoring to acquire her training license for over a year but had, she said, been repeatedly, expensively, and needlessly obstructed by the TMB on account of two technicalities, one being a bout of depression that she had suffered several years earlier, the other being an arrest for an unpaid speeding ticket when she was a teenager.

Fuentes spoke for over fifteen minutes—twice the limit—and the details of her story were less striking than the Kafkaesque way they followed each other, with increasing absurdity, a litany of bureaucratic provocations which appeared all the more grotesque when compared to the forthright way they were related, with zero expectation of justice. By the time she was done the entire tone of the hearing had changed.

"If I am retaliated against because of my testimony, then so be it," Fuentes concluded, "but I will sleep at night knowing that I have done what is right."

Her voice, speaking for all doctors, was like a bright reminder of what truth sounded like, and its extraordinary power to cut through any amount of bureaucratic obfuscation. When the applause died down Representative Debbie Riddle was the first to respond.

"First of all I want to thank you for your courage," she said. "Sam Houston once said, and I had this hanging up in my house while my children were growing up, he said, 'Do right and risk the consequences.'"

"Yes ma'am," said Fuentes.

Riddle then asked for a clarification as to when exactly the speeding ticket had occurred.

"It was twenty years ago," Fuentes said.

"I'm just curious," Riddle went on, "in the past twenty years has anybody in this committee hearing room ever not had a ticket? . . . I mean if you've not had a ticket in twenty years would you raise your hand? Okay, well it looks like we've got a few."

"In the defense of the board," Fuentes put in, "which is going to be ironic but, in their defense it was an arrest that resulted from a speeding ticket and not just the speeding ticket."

"My daughter had a similar situation when she was in law school," Riddle remarked.

"Right," said Fuentes. "Well I gave it to my father. It got put on the kitchen table and when it disappeared he thought I paid it and I thought he paid it."

It was around here, with the mention of the kitchen table, that the Regulatory Subcommittee of the Texas House Appropriations Committee slipped the bonds of bureaucratic convention and turned into something markedly more human. The effect was startling—also liberating, as could be seen by the jocularity of the comments that followed.

"I just want to say to Representative Riddle," an unidentified woman said, referring to those in the audience who had raised their hands, "that this goes to show you we do have some perfect employees."

When the burst of laughter had passed the chairman cut in with a football reference (this was, after all, Texas), presumably in response to a new arrival: "Members, down to my left here is the new Chairman of

Sunset [Committee] and I just want you to know that I didn't think it was very Christian what your school did to Texas A&M two weeks ago."

"I beg to differ," the other chairman replied. "That was the most Christian thing we could do."

More laughter. And you could feel the contraband humanity filtering back into the room.

"Ms. Fuentes, we're proud of you for coming today," the chairman resumed. "It takes a lot of courage and we will ask the Board to keep us apprised of your future situation and hopefully it can be resolved soon."

Fuentes was followed by a flood of other alternative medicine doctors echoing the charges that had been made by Rea in his letter to his patients. One spoke of his patients being denied the right to testify on his behalf. Another spoke of being prohibited from recording his hearing or taking notes. The counsel for the Association of Physicians and Surgeons, Andrew Schlafly, reported to the subcommittee that his organization considered the TMB to be the worst medical board in the country.

As doctor after doctor got up to tell their story you began to get the sense that what had been committed was a crime of occlusion. It was the same sort of complaint that sensitives made of most doctors—the same phenomenon only one tier higher—a complaint about not being *seen*. Which seemed to suggest that maybe the hierarchy was the problem, that wherever there was hierarchy there would always be this same narrowing and occlusion. Until a real, human voice cut through the nonsense to remind everyone of what was being left out.

"They're condescending, they're harassing, they're abusive," one osteopath said. "They're really hard on doctors and good guys. I'm an Eagle Scout, darn it. I don't deserve that."

And suddenly you found yourself thumping the table in agreement.

The osteopath had been accused by the TMB of treating a patient with "vodka drops." He dismissed the charge as ridiculous but he wanted to make one thing perfectly clear.

"I'm a family doctor but I do [have some] pretty tough patients who are really sick. This [one] little old lady had burning leg pain she couldn't get rid of for five years. She saw twenty-two doctors. Nobody could help her do it. I came up with a good idea that helped her and it wasn't vodka drops, though if it would have helped her I may have used it, and I'm serious about that, I may have."

And this was the point that the doctors kept trying to emphasize: their profound dedication to their patients, beside which even the noble conventions of medicine took on a secondary importance, and about which the functionaries of the Texas Medical Board clearly knew zilch.

"I am very passionate about my patients," the next doctor said, "and I take pride in that I listen to my patients. My average appointment is an hour. My average new patient appointment is four hours . . . Yes, and that's why I don't take insurance payments or assignments because I tend to spend a lot of time with my patients. They have usually been to see anywhere from twenty to thirty doctors before they end up in my office. I'm kind of like this previous doctor, I'm kind of a last resort for them."

At what was surely the climax of the hearing, about nine hours in, a doctor speaking for a group called Texans for Patients' and Physicians' Rights (TPPR) demanded an end to the practice by which complaints could be made against physicians anonymously, such that they could not confront their accusers.

"No more star-chamber proceedings held in secret," the TPPR rep declaimed. "No more prohibition of notes and recordings in hearings. Don't tell me, Dr. Kalafut [Roberta Kalafut, president of the TMB] that you can take notes. You were in my hearing when you intimidated my personal assistant and told her to shut off her computer. You stood up and made an ass out of yourself. It was horrible. If I hadn't been in such a precarious situation I would have given you a good tongue lashing. You deserved it. Your momma needed to take you over her knee is what she needed to do."

At this point, Kalafut stomped out of the room in a fury.

By the time Rea himself took the stand, his complaints, largely reprising what he had already written to his patients, seemed almost mild by comparison. All the same, the committee took his comments to heart. "First of all, Dr. Rea," Representative Debbie Riddle said, "I think that from everything that I've heard about you and from everything that I know, quite honestly I think it's a privilege to have you as a physician here in our state."

She went on, referring to the Texas Medical Board staff seated behind Rea: "But one of the things that is making me absolutely furious and I'm just going to say it right here is the people behind you and the expressions on their faces of such arrogance while a man of your character and of your accomplishments and your dignity is sitting there testifying. You folks need to be ashamed."

Riddle's remarks were met by huge, cathartic applause, and, even reading them, you could feel a melting gratitude to see the buck finally come to a screeching stop. Clearly, the TMB had been at fault, not Rea, and they had received their due. The doctors had spoken, and they had been heard.

Several months later the executive director of the TMB resigned, followed shortly thereafter by the president of the board. The TMB complaint against Rea was eventually whittled down to an order that he provide patients undergoing antigen therapy with a revised informed consent form saying that the therapy was not FDA approved and that its effectiveness was disputed. One year later, the Texas legislature passed a law prohibiting anonymous complaints to the TMB.

And so justice was finally served, and Rea was allowed to keep practicing. Certainly, his methods may have been unconventional, but then so were his patients. For them, he was a doctor of last resort. Therefore punishing Rea for being unconventional was essentially just a roundabout way of punishing his patients. Clearly, the *Nightline* report had it backwards: if anyone was guilty, it wasn't Rea but the Texas Medical Board.

Except . . . I later learned that the American Association of Physicians and Surgeons, the highly vocal group that seemed to be the organizing force behind those who spoke out against the TMB, was actually a fringe advocacy group primarily dedicated to "fight[ing] the government takeover of medicine" and defending "doctors who have been mugged by Medicare," and more widely known for opposing abortion, denying human involvement in climate change, denying that HIV causes AIDS, defending Rush Limbaugh, and claiming that Obama used "neuro-linguistic programming" to hypnotize people into voting for him.

Andrew Schlafly, counsel for the AAPS, was none other than the son of Phyllis Schlafly, the conservative political activist notorious for her successful campaign against the Equal Rights Amendment in the 1970s. Schlafly junior was also the founder of Conservapedia, the ultraconservative, Christian alternative to Wikipedia that claimed Shakespeare as a political conservative (based on, among other things, the pro-family message of *King Lear* and the anti-tax message of *Coriolanus*), and went on to compare him, William Shakespeare, unfavorably to Phyllis Schlafly, who after all was four times more prolific.

As for the guy who suggested that the board president's momma needed to take her over her knee, he turned out to be a Houston talk radio host who had claimed that if same-sex marriage was legalized "kids will be encouraged to practice sodomy in kindergarten," and that the gay rights movement was backed by "satanic cults."

That such figures as these were coming to the aid of alternative medicine doctors could only be explained as one of those political glitches wherein the right, in a kind of mad Möbius backbend, reached so far around that it actually reconnected on the other side with the left.

What's more, a broader, more historical analysis of the Texas Medical Board's disciplinary zeal revealed that it actually began five years earlier, when tort reform was passed, limiting the damages that could be awarded in medical malpractice cases. At the time, patient advocate organizations interpreted this as decreasing physician accountability, which is why the TMB began ramping up their oversight, by way of compensation.

"During my Senate confirmation hearings," Kalafut later told the *Houston Press*, "I came away with the impression that if the Legislature was going to pass tort reform, which the physicians wanted, we at the Medical Board would have to step up to the plate and regulate our own."

At the end of the day there was no single truth awaiting gemlike in the mire. Everyone had an angle. Rea. Fuentes. Kalafut. Schlafly. It had never been a simple choice between kookdom or corruption. The forces of both were always at work to one degree or another, forever inveighing against each other in the same eternal dialectic.

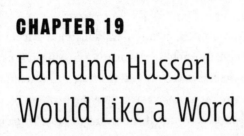

CHAPTER 19

Edmund Husserl Would Like a Word

He lies there waiting for full dark to fall, the wind tearing at the tent in 50-mile-an-hour gusts. Hopefully it'll hold. It's just a cheap Walmart tent, but on the plus side it only takes a week to off-gas, and besides, the sun's rough tongue shreds pretty much anything that casts shade inside a month or two, so it's hardly worth spending more.

These are the difficult hours, between 7 and 9 p.m., when you just have to lie there and wait for time to conquer itself. At least the heat has begun to fade. During the day it can reach 115 degrees. There's a big difference between 95 and 105, and 105 and 115, by the way. It's not exactly linear, and even at 95 you still have to shift the 50-pound water jugs around the empty house to keep them in the shade. The water here is some of the worst in America, toxic with metals leached up from the ground. Even shower water must be hauled in. No toilets either. Poop goes in a hole. After three years of this you get over the stigma.

Reading helps pass the time. Trollope, Eliot, James, even George Gissing. Discovering Gissing was a lifesaver. Thousands of pages of hospitable literature, all available online for free. It was enough to make you feel thankful, despite everything.

But you can't bring the iPad into the tent at night because unlike your phone there's nowhere to hang it, and what if there's a flood? And you can only read for half an hour or so on your phone before your vision starts to blur.

So you lie there in the windy dark sucking a Ricola and YouTubing Schubert. Between the Schubert and the Gissing it's as if you've been plunged back in time. Evicted not just from civilization but modernity itself.

Which actually matters less than one might think. What matters is to feel these deep parts of yourself, ignored by everything else, being recognized and spoken to. It's not as good as being heard, but it's close.

Night comes at last. And now the only light is from the flashlight under the car, to keep the packrats from gnawing on the gas line and building nests in the engine. Your mind drifts to a time not too long ago when you went out for dinner with friends at Il Buco, an Italian place on Great Jones Street in New York. There was octopus with roasted artichokes and lemon aioli, and rabbit ragu with sage and kabocha squash, and trout, braised short ribs, and wine that rang pure as a church bell. Oh God, for a meal like that. When you're surviving on dried fruit and granola it's best not to think about it. It's like a limb has been taken from you. Like a part of you is just gone.

———

From the south a rumble of thunder. David Reeves looked up. Faded black jeans, horn-rim glasses, white sneaks. Sunburned nose and a boy's short, ruffed hair. Except for the nose he looked very much out of place. The kind of young intellectual you might find chatting with a friend some Sunday afternoon on one of those thickly painted green benches in Central Park. It was easy to see him, fingers in pockets, slapping down a New York sidewalk. As I often had when I lived there.

"Gonna get some rain pretty soon," he said. "Monsoon season out here basically means every day it rains at some point. It's like being out at sea. You see it coming in."

This was true. Even as he spoke a gray parallelogram was shouldering over the Empire Mountains and raking the land to the south. The remote acreage on which David had made his home for the past three years was elevated slightly above the scrubland, and to the south especially you could see for miles. It felt about right, a more fully realized version of the sort of place you find yourself visualizing in slow-moving checkout lines, overcrowded subway cars, unquiet libraries. A place where you found yourself restored to a full 360 degrees of landscape, exposed only to the grit of the wind and the indifferent roll of the earth. The spaciousness seemed to go on forever, inviting a repose to match its own. And from this repose there arose a quiet expectancy, a waiting that had been there all along. Waiting for what? Predators? Weather? The crashing return of all that was absent? The mountains remained where they were, not encroaching, hour after hour. It put you in the mind of that line by Rilke, about the beauty that serenely disdains to destroy.

We'd been talking in the dirt driveway, J and I on metal folding chairs, David on a cinder block. The house was a simple one-story affair with a metal roof and a stucco exterior and it was supposed to be EI-friendly but two years ago the pipes had burst and with the mold that followed David hadn't been back inside since.

Like so many others I'd talked to David had been blindsided by EI— perfectly healthy for forty-five years and then suddenly, a few days after they started spraying his apartment building for bedbugs—

"I woke up in the middle of the night, like, what the hell just happened? I just felt like I had to go to the ER. It literally felt like my immune system just went berserk."

Looking back, David remembered living in a very moldy dorm for several years in college. And his New York apartment was twenty-five blocks from Ground Zero. ("I did a lot of walking in lower Manhattan post-9/11," he said.) Plus, the Flatiron Building where he worked was over a century old, and known for being a dirty building. So maybe Miller was right, and David's pesticide exposure was simply the last in a chain of such incidents that pushed his body past its limit.

Six months later he was moving back in with his parents in Connecticut, only to find that he couldn't tolerate the mold.

"I was just sick all the time," he said. "Crazy reactions. Like muscle twitching, nausea. I could barely walk."

David went to see top doctors at the Occupational & Environmental Medicine program at the Yale School of Medicine—but no one had anything helpful to say.

"It was astonishing how dismissive and just flat-out ignorant they were," David said. "It's like the nineteenth century. It really is. Y'know, you read in the *Times* about cancer research over the last twenty-five years, like immunotherapy, and, man, there is *nothing*. This is like the Dark Ages for EI."

The doctors at Yale didn't see it that way. But such was the nature of experts. Harder even to believe than EI was that they could know so much about one thing and nothing whatsoever about something else. In their view knowledge grew evenly, like hair, when in fact this sort of patchiness—like prehistoric stoplights in an age of self-driving cars, or religious fundamentalism in an age of civilian spaceflight—was one of the defining quirks of modernity, or postmodernity, or whatever they were calling it these days.

Eventually David quit his job and, having heard about Snowflake, bought a one-way ticket to Arizona. He didn't end up moving there, though.

"It's a strange community," he said. "There's all this jealousy and infighting. And the people there really seem almost obsessed with toxicity. You meet some of these EI people and it's, like, shocking. They'll come up to you without even knowing your name and just start telling you all their symptoms. There's something called the HEAL group out here, and they do a once-a-month meeting. But I stopped going, because, I dunno, man. Like, cancer victims don't deal with it this way."

It was interesting to hear this from a sensitive. Unlike so many others I'd spoken to, David had somehow managed to resist EI's tribal allure. Nor did he come across as cynical or resigned. On the contrary, he was surprisingly upbeat.

"When you're sick," he said, "this is all you have, man. This is your life. You don't get many more days. And at the end of the tally you don't sit there and say, 'Y'know, for half of my life I felt kinda bad.' This is it, man. Like, okay, you don't feel great, you're having some kind of problem. But my personal way to get through it is to try to live some kind of life, and *do* something, and grow intellectually if I can, and not stagnate."

I appreciated this about David. The disease didn't seem to own him like it owned others. It might even have helped emancipate him by releasing him from the New York bubble. When he first moved out here, for instance, he was shocked to see open-carry weapons in the supermarket, and everyone wearing cowboy hats and driving pickups. Then one day, bored and lonely, he dropped down to Mexico. He got as far as Naco, a little drug-running town just over the border. After ten minutes there, with heavily armed Mexican paramilitary shouting at him and searching his car, he was ready to turn around and come back.

After that, he said, he never felt more connected to other Americans — regardless of their political views.

"And I think that's a really important distinction," he said. "A lot of people don't get outside America, don't go to places where there really is crime and just utter poverty. If you went to a favela in Rio and you got back to Arizona or Connecticut or wherever you'd be, like, 'Thank God I'm back in America.' It wouldn't matter who was president. The differences become this selling point for the media. I could go on for hours about it, but I think there is a fundamental American sensibility. We're all part of this capitalist system, we all have bank accounts, we all have Internet, we all watch HBO. But it's like you have a pistol and I don't have a pistol and suddenly we're archenemies and have to go hang our political leaders. What is this? I dunno."

David's desert exile offered this much: perspective on the folly of tribalism, whether among sensitives or more broadly among various political factions. At a time when everyone else was growing ever more paranoid and warlike, Democrats vilifying Republicans, Republicans vilifying scientists, scientists vilifying doctors, doctors vilifying sensitives, sensitives vilifying chemical companies, government regulators, the media, each other, David was ready to do the exact opposite — and not in some mushy, bleeding-heart way but on the actually pretty conservative basis of our

common identity as Americans. Because our identity as Americans preceded everything—including sickness.

It was a nice idea. A really nice idea. Surprisingly old-fashioned and
idealistic yet also somehow pragmatic, solution-oriented, the way Americans used to be.

I liked this about David. The nerdy way he sat there on a cinder block
at the end of the world, entertaining old-fashioned ideas. I liked, too, his
warmth and openness and willingness to profess mystification. Even the
way he said my name, as if we'd known each other for years. Like, "Oliver,
you don't want to get sick, man. It's just a pain in the ass."

When he said this, I believed him. More than any other sensitive I'd
talked to. More than the scientists, the doctors, more, certainly, than the
chemical companies and government regulators. Which was important because it suggested that maybe the best evidence for the reality of his EI
was not studies or blood tests but character. There was nothing, after all,
defensive about David, as there had been with Scott. Nothing jittery or obsessive. He seemed completely at home with himself, as well-adjusted a person as one could hope to be in his circumstances. EI was a nuisance the way
asthma was a nuisance. It had nothing to do with him as a person. His words
and behavior betrayed not the faintest trace of complicity with the disease.

Despite the physical challenges, in other words, David was still very
much David. Even here, amidst the rattle of the wind and the snakes,
he was literary and urbane. Even here, cruelly divorced from the city
which had given him so much of his identity, his identity remained intact. And to the extent that skeptics were always trying to frame EI as
"merely psychological"—an identity issue—this carried a certain amount
of weight. If EI really was merely psychological it somehow managed to
be so—and to have a truly devastating impact in the process—in an unusually compartmentalized way, without compromising any other aspect
of the sufferer's identity—even after this identity had been stripped of all
that once affirmed it. It would mean that the true psychological origin of
the EI was split so cleanly from the rest of the self as to resemble a kind
of schizophrenia.

And how likely was that? Could the same be said for anorexia? Or

PTSD? Granted, one couldn't rule out the slim possibility that there really was an entirely different side to David, and that I was just being gulled by familiarity. But the most compelling basis for this scenario, it seemed to me, was Claudia Miller's inverted addiction model (addicts, after all, were known for their radical, chemically induced shifts in character—and as a population were no less heterogeneous), and even addiction couldn't be dismissed as "merely psychological."

So if you still insisted on describing EI this way, as "merely psychological," as if EI were some kind of epic conversion disorder, then at the very least our whole notion of what was meant by "psychological" was due for some serious rethinking.

Perhaps the challenge of parsing EI derived from a faulty intellectual framework. For somehow it always turned into this hard binary, psychological or biological, and the violence of the opposition always seemed to cause some sort of trauma. The field of allergy itself emerged in exactly this way, such that you might even regard the EI controversy as a compulsive reenactment. At the time, around seventy years ago, the controversy concerned a different unknown: food allergy. Most other allergies could be reliably detected with a skin test, but for some reason food allergy could not. So the question became whether food allergy should be regarded as "real."

For allergists at the time, the question once again came down to how you responded to uncertainty. Those who regarded uncertainty as a threat to their profession (at the time, the field of allergy was still struggling for legitimacy—not to mention funding) opted for the more rigorous definition of allergy, dismissing food allergy as psychosomatic and renouncing those who treated it as quacks. (To be fair, defending their profession was, arguably, the best way to ensure that they could keep treating those patients that they knew for a fact were sick.) Those less troubled by uncertainty opted for the more inclusive definition of allergy—the one originally put forth by Clemens von Pirquet in 1906 as simply "any form of altered biological reactivity."

Theron Randolph, the granddaddy of environmental medicine, belonged to the latter group. True, he couldn't objectively verify the presence of food allergy. But there were other things he could do.

"I simply listened," he said. "I listened to my patients."

Randolph was one of those docs who spent as many as four hours on a single intake.

"It occurred to me that if I recorded only what I knew from past experience and training to be important," he later said, "I might be equally misinformed 10 years later."

Just as Hope, Rea, and Bernhoft and so many other EI docs began as patients, so did Randolph. Among other things, he was allergic to peanuts, maple, and corn. His wife—his second one, anyway—was highly allergic as well. Depending on how you looked at it, this either biased his perceptions or afforded him an unusual opportunity for insight. For, like Gray, Randolph was foremost a clinician. It was for this that, just as Gray had been booted from his post at the Arizona Division of Occupational Safety and Health, Randolph was booted from his post at Northwestern University—excommunicated, basically, branded "a pernicious influence for medical students," as if Randolph himself were some sort of allergen and the fledgling field of allergy the sensitized body.

The hostility and sheer endurance of the binary framework—dating at least as far back as Descartes—was striking. In some realms, the psychological-biological rift was being challenged—in chronic pain research for instance. Yet if EI revealed anything about mainstream medicine it was the degree to which the Cartesian bias persisted. Indeed, it was built into how the whole field was structured—by specialty, around individual organs, thus steering the physician's attention away from the patient's experience of disease in favor of whatever biological component could be identified as the problem. In this view, the patient herself was (like a drug addict) an obstacle to healing—not a contributor. The rise of evidence-based medicine only reinforced this trend, displacing the experience of individual patients with the generalized results of clinical trials that "on average" suggested the best course of action.

Meanwhile, whenever evidence surfaced to challenge the Cartesian paradigm it could always be neatly dismissed as "placebo"—a mere spoiler in the ongoing grind of randomized double-blind studies, an exception to an otherwise thoroughly sensible and efficient system. Never mind that the domain of the placebo, encompassing a broad range of biological, psychological, and contextual variables, could easily be a field of medicine unto itself. Only ignorance allowed the charade to continue—an ignorance regularly betrayed by the way so many scholarly articles on placebo ended, with a kind of helpless throwing up of the hands: *additional research is required.*

After reading enough of these papers, however, you got the feeling that it wasn't new research that was required so much as an entirely new mind-body framework, one that could at least accommodate the placebo effect without roping it off as a pesky exception. Which offered some explanation for why the Cartesian bias persisted. Because it was easy. It gave physicians a way to keep doing their job without having to (1) single-handedly solve the mind-body dilemma and then (2) find some way to integrate the solution into an impossibly ramshackle health care system.

The difficulty of the second problem made it easy to forget that the first had actually already been solved. By the Germans, naturally. But the solution to the first problem intimated why the second would always be such a challenge.

———

There's no such thing as consciousness in the abstract. So said Edmund Husserl, the German phenomenologist, in 1913. All consciousness is directed at something—an ice cream cone, a political candidate, a candle— and this interface of candle and consciousness precedes any meaning that we might later choose to append. Meaning, in other words, commences with the thing itself. It is not added later.

This shift in focus, from some idealized point of self origin to the messy interface with reality where consciousness emerged, constituted a major check to the fatheadedness of Descartes, who took the primacy of his own intellect for granted. Husserl's work came as a reminder that, whatever our philosophical aspirations, or professional pretensions, we were all a part of the world, and indeed as creatures constituted a gloriously emergent, hopelessly entangled processing of it. This was something that scientists, with their posture of stately objectivity, loved to overlook. As well as all those who relied on the authority of science for their expertise. Like doctors, whose immaculate examination rooms and featureless scrubs seemed expressly designed to enable the view that medicine required the elimination of all worldly context, including the context provided by a patient's subjective experience.

The overthrow of Descartes continued with Husserl's followers, mainly Martin Heidegger and Maurice Merleau-Ponty. Heidegger did away with the last, old-school, chin-on-fist posturing of Husserl, allowing the humble truth about human Being to finally come into view. The "primary issue is never to provide an explanation," Heidegger chided, "but rather to remain attentive to the *phenomenon* one seeks to explain—to what it is and how it is." Which was basically just a nice way of pointing out philosophy's eternal, anxious bias, like a two-year-old always yammering *why* instead of shutting up long enough to wonder *how*, and *what*.

In refuting Descartes Heidegger challenged the ascendant influence of Freud, who had concluded that identity, like Descartes' "existence," was a riddle to be figured out, and was eagerly aping the reductive tendencies of his more hard-science-based colleagues to this end. What emerged under Heidegger was a far more ecological view in which human Being flowed into the world and was thoroughly permeated by it. In his telling, specificity replaced abstraction, messy actuality the Platonic ideal. "What we 'first' hear," for instance, he wrote, "is never noises or complexes of sounds but the creaking wagon, the motorcycle. We hear the column on the march, the north wind, the woodpecker tapping, the fire crackling. . . . It requires a very artificial and complicated frame of mind to 'hear' a 'pure noise.'"

Nothing was "pure." Nothing was "textbook." The apartness between subject and object that language struggled to assert by way of satisfying our need for agency and control collapsed in Heidegger's account and was replaced by an entirely new, semi-penetrable language of his own designed to capture human beings as they really were, neither noun nor verb but a kind of tingling fusion, an ongoing impromptu collusion with reality of the sort that Joyce, beating Heidegger by five years, sought to capture in *Ulysses*. Human beings were not, as Byron liked to complain, in swaggering self-consolation, half-deity, half-dust. The existential melo-drama only evidenced the lousy tools we used to make sense of ourselves, not who we actually were—*how* we actually were. In truth we grew into the world from birth as a plant shoots roots into the soil until speaking of ourselves as in any way separate or apart no longer made sense.

It was as close as anyone had ever come to offering an objective view of subjective existence. And what followed from it was an altered sense of the regard we owed each other, as humans. The implications for medicine were obvious, as Heidegger well understood.

"There is the highest need for doctors who think," he said, "and who do not wish to leave the field entirely to scientific technicians."

In 1959, at the invitation of a Zurich psychiatrist, Heidegger agreed to give a series of seminars explaining the implications of his philosophy for medicine. His only conditions were: (1) no philosophers allowed—medical professionals only; and (2) no vexing questions about his Nazi past.

The seminars continued over the next ten years as Heidegger strug-gled to bridge the gap between theory and practice. The first seminar began with Heidegger chalking a semicircle (upright like an ear, not recumbent like a smile) on the blackboard to represent the patient's fundamental openness to the world and everything in it. The patient, he explained, was not an object, and the reductive language of scientific medicine—Pall's language—could never fully account for the patient experience. How, for instance, was scientific medicine to make sense of grief? he asked. Could you, for instance, with a sufficiently powerful microscope, identify "fare-well molecules"? In Heidegger's view, scientific medicine couldn't even make sense of something as simple as blushing. A person might blush for a handful of reasons: shame, fever, or perhaps on entering a warm chalet after a cold evening walk.

Heidegger was followed by Maurice Merleau-Ponty, who grounded human Being even more fully in the physical body. "I do not *have* a body," Merleau-Ponty would say, "I *am* my body." Which was not a glorification of the flesh so much as an acceptance that "there is no inner man." "Man," Merleau-Ponty said, "is in the world, and only in the world does he know himself."

This was in a way a great relief to hear because (as those who came after Merleau-Ponty later argued) humans would never be replaced by AI: because even if machines could match our cognitive processing power they would never be immersed in the world in quite the same way. Absent this immersion, not even morality made sense—as recent experiments had shown. Committing violence in a nonimmersive virtual reality, for instance, left no mark on the psyche whatsoever. The same violence committed in an immersive virtual reality, however, could cause lasting psychological trauma.

In solving the mind-body problem, the phenomenologists brought us considerably closer to understanding human existence as it really was, and therefore who the patient really was, and thereby seemed to offer a far more sensible grounding for the field of medicine. If the patient was not simply a mechanism of flesh, an animate corpse awaiting the scalpel, but rather an arc opening to and in a sense one with the world, then this told us something about who the doctor needed to be to treat her: not just a cutter.

And yet Heidegger's and Merleau-Ponty's ideas barely penetrated the world of biomedicine. To the extent that they did it was mostly in the form of chiding editorials that appeared in medical journals every few years enjoining doctors "to remember that our patients are people," and to "embrace a compassionate attitude that shares existentially—not simply technically—with a patient's suffering," and so on.

You could say that doctors' lack of interest in the phenomenologists' ideas was an attentional issue. Humans were problem-solving animals and solving problems required focus. Focusing on two different things at once—mind and body—impaired the quality of that focus. You might as well ask Heidegger to also cut hair. In this sense the reductivism of medicine simply mirrored the nature of human attention. Secretly, on some level, every doctor has at some point wished they were a veterinarian. Veterinarians didn't have to worry about the individuality of their patient. Certainly not to the same degree. Dogs were dogs. This was, by the way, the beauty of dogs. They had just enough individuality to be real and not one iota more. In this respect, dogs were what every doctor wished their patient to be, just as food companies wished the same of their customers. The only difference being that doctors, being human themselves, were presumably conflicted.

Not surprisingly, the field that took the greatest notice of Heidegger's contribution was psychiatry. To psychiatrists, the notion that patients deserved attention not just as bodies but as people was not exactly unfamiliar. Besides, Heidegger's ideas offered a timely hedge against the waning influence of Freud, shifting attention away from "drives" as the source of behavior and toward a more contextualized understanding of "affect."

Within medicine, meanwhile, Heidegger's ideas received the warmest reception from nurses, who, like the shrinks, hardly needed reminding that reductive medicine left a lot to be desired. Their entire profession was premised on making up the difference. It was also 95 percent female, and the disproportion only reinforced the miserable way the Cartesian split had always been allocated between the genders: the doctors did the cutting; the nurses did the caring. And never the twain shall meet. Indeed, studies had shown that communication between doctors and nurses was terrible, with doctors ignoring their nurses much as they often ignored their patients.

As for Merleau-Ponty, his insights on the nature of embodiment contributed significantly to the now popular theory of embodied cognition, and have been increasingly borne out by advances in neuroscience. (Using fMRI, for instance, researchers have shown how action verbs like "kick" and "lick" activate the same areas of the brain as the movements the words refer to, thus suggesting that semantic understanding is at least partly grounded in physical experience.) But little of this has informed how medicine is actually practiced. Possibly it could have gone otherwise if M-P had had a chance to complete his work on embodiment or pursue its implications. Regrettably, he dropped dead at age fifty-three while preparing for a class on Descartes, who thereby had the last laugh.

Chances are it wouldn't have made a difference anyway. The phenomenologists' descriptions may have been accurate. But the truth was not what was wanted. Because the truth, it appeared, was not practical.

CHAPTER 20

What You Can Learn from a Butterfly

G ray paint strokes of rain advanced from the south, along with the howl of a freight train carrying westward on its glinting ribs the day's leftover light. A hawk kited above the bedclothes rumple of the hills. The wind stirred, uneasy. J munched coconut chips.

It wasn't much, this plot of land, but David was attached to it. For him, it was home.

"Even though that's total illusion," he said, "it's a wonderful illusion. Home is what you can go back to, and know what to expect."

He still ached for New York, and all that had been taken from him. But he had said his goodbyes, and made a kind of peace.

"Look, man," he said, "I may never get back to New York, but I had it for twenty-five years, and was there at this incredibly exciting, vibrant time, from 1989 to 2000-whenever. And it was amazing. And not everyone gets that. Not everyone gets Paris and not everyone gets Rome. So I got it, and I had it, and the memories become the good times."

We left him feeling sort of mournful. It was like saying goodbye to someone who had already gone, like talking to a figure from the past, or a living statue that would revert to rock the moment you turned away.

J pointed the Rover west toward Tucson and the next Whole Foods. The plan was to eat there and tack north, barreling through Gray's toxic zone to reach Sedona, where J had a house.

It had been over a week without any word from Brian. We'd been leaving messages for him and checking the Facebook groups daily for up-dates. The online community was in a minor uproar.

"Assuming you are off-grid. Let us know you are ok when you can!" one person said.

"Not sure what his location is," someone else said, "but . . . if he needs help getting to emf free location we can work on it. Brian has struggled too long . . . My heart breaks and I miss his narratives."

"I had been thinking it had been a while since any Brian reports too, sending hugs."

"Someone find Brian I can't sleep . . ."

Rain tickled the windshield. Lightning jumped to the southwest and the Rincon range withdrew behind a gray veil. We thought of David in his tent. The tickle became a clatter—a breath of silence beneath an overpass—the clatter again. We sheltered behind the rectilinear bulk of an eighteen-wheeler toiling westward on the bright wet road.

"I started feeling like crap a little while ago, up there," J said. "It comes in these waves. I feel fine now, I can talk. Sometimes it feels like I can't talk."

David had mentioned maybe one day buying ten acres of his own, on land bordering protected wilderness, and building a simple house. Now J roused himself and began plotting how he might do it. "Concrete floor, single slope roof, all steel framing . . . I could put a solar array on top. If that wouldn't fry my head with EMFs."

He lapsed into silence again, his show of industrious pragmatism fad-ing. The mournful tones of Radiohead's "Reckoner" came on the playlist.

"What if Gray is right," James said, in a very different voice, quieter now, as if he were alone in the car and talking to himself. "What if I never get back to normal. That's one thing that's kind of been bothering me.

I haven't been feeling good today but . . . You heard Dr. Gray. He said, 'You'll always have the sensitivities . . .'"

I held still, listening. It was the lowest I'd ever seen him. And suddenly it occurred to me that maybe there was a connection between the physical symptoms of EI and his emotional state. Like maybe the real reason he wasn't feeling so hot was not because his EI was acting up but because Gray had told him he was fucked. In other words, maybe for James EI really was just an epic conversion disorder. His body's expression of an inability to process anger or grief.

Dr. Bessel van der Kolk had made a strong case for this phenomenon in his 2014 book, *The Body Keeps the Score*. In it, he tells the story of one woman who, by way of protecting herself from emotional anguish, had learned to transmute it into an asthma attack—which by comparison was far more manageable. Other patients experienced migraines, chronic back and neck pain, fibromyalgia, digestive problems, and chronic fatigue.

"They learned to shut down their once overwhelming emotions," van der Kolk wrote, "and, as a result, they no longer recognized what they were feeling."

It was a variation of the PTSD argument, with the difference that van der Kolk believed that trauma could be inflicted even in the absence of explicit abuse. In children in particular, simple neglect was enough.

Arguably, the callous way James's father responded to the suffering caused by James's childhood allergies ("Tough it out") amounted to a kind of neglect. And from a shrinky remove that dynamic seemed the obvious template for the sensitive's relationship to the doctor. The parent after all is the original expert, and the expert's imprimatur is what sensitives always seemed to want.

The more I thought about it the more possible it seemed, especially taking into account a story James had told me about crashing his motorcycle. Fifteen at the time, en route to school, he was leaning into a curve going around 90 when an inattentive truck driver let his rig drift over the yellow lines. James went off the road, tried to steer the bike back on, but the front wheel slipped on the road edge and the bike pinwheeled. James surfed the pavement on his left side, shattering his helmet visor and losing skin from ankle to elbow.

"The sleeve of my leather jacket was pushed all the way up around my arm," James said, "and the skin was taken off in a chunk at my elbow. The left side of my jeans was completely ripped off. It looked like I had on Daisy Dukes on that side. Blood and pavement burns, sand and grit and asphalt in my skin."

It must have been terrifying. "But I didn't cry," James said, as if this were an accomplishment.

But for him maybe it was. He toughed it out. Didn't even wait around for the ambulance but asked the truck driver to give him a ride home. On coming through the door he received no words of comfort or concern from his father. Instead his father simply told him to rinse his wounds with peroxide and then sent him off to school.

If this could happen, if a father could send his son to school after a terrifying motorcycle accident, bleeding beneath his clothes, what else might have happened that James couldn't talk about? And how much of a difference was there, really, between a father who shows no sympathy when his son is hurt and a father who actually does the hurting himself? In both cases the message was the same: Your pain doesn't matter. *You* don't matter.

Certainly the PTSD diagnosis would have explained a few things. Why he was stuck in a pattern of constant flight, for one thing. Why he always seemed to be covering for his father's callousness: "I don't know, maybe he wanted me to be stronger."

It made me think again of what James had said the day I met him, about benching 365 when he was still working for his father. The only reason he stopped was because his joints began to give out. All that power—like Hof's over-physical heroics—screening him from grief.

"Children are . . . programmed to be fundamentally loyal to their caretakers," van der Kolk had written, "even if they are abused by them. Terror increases the need for attachment."

Possibly, then, EI was just a way for James to gather all his denied suffering into a single package—and then assign it to some L.A. quack to resolve.

It was a neat idea, but I still couldn't quite believe it. EI was too complex. At times it seemed to me like a symbiote, evilly threading itself along its host's finest wires until it became indistinguishable from him, as difficult to excise as it was to excise mind from body or body from mind.

Perhaps it was better to discard the mind-body paradigm altogether and start again from the beginning.

What was needed was something more than philosophy. More than a mere description of how mind and body came together at the phenomenological verge. What was needed was a way to bridge the two, to forge a coherent, scientifically sound path from one to the other.

On the one hand the need for such a path had never been more obvious or urgent. On the other, a path such as this had never existed before in Western medicine, and there was little reason to think one could be invented now. In fact, you'd have to be either a genius or a fool to even try.

This was the story of George Engel.

—

Engel didn't start out with the goal of transforming medical care. In fact he began his career in the 1940s as a hard-core biomedicine guy. Raised in a house full of books, he had a deep-seated respect for knowledge — and for *causes*. Sherlock Holmes was a role model, as was his uncle Manny, a renowned diagnostician who bore a striking resemblance to the great detective. Uncle Manny once diagnosed Gershwin with a brain tumor after conversing with him briefly at a party where Gershwin was playing piano. Another time, he diagnosed President Harding with heart disease after someone described Harding greeting guests at the White House in "the Napoleonic pose." Tell Coolidge to get ready to be president, Uncle Manny said. A few weeks later Coolidge was sworn in.

Engel was trained on the dead body, not the living one. One summer he performed over three hundred autopsies. Thankfully, Engel was also a fainter, which stirred his interest in the phenomenon and later led him to study with a man known for his research in this area, the eponysterical Soma Weiss. It was Weiss who teamed Engel with John Romano, a young psychiatrist interested in neurological disorders. Engel was displeased to be working with a psychiatrist, but his views underwent a profound shift when, on hospital rounds with Romano one day, he observed the way he pulled up a chair at a patient's bedside and invited him to tell his story. One didn't approach corpses in this fashion. Shockingly, Romano's methods yielded useful data.

In the years to come Engel became increasingly interested in the patient as something more than a working corpse. He went on to become a leading advocate of psychosomatic medicine, eventually ending up at the University of Rochester, where, with Romano's help, he assembled an innovative curriculum that integrated the insight of psychiatry with the practice of medicine. Engel became known for tutoring his students to look beyond the body for clues—whether any get well cards were taped to the wall by the patient's bed, for instance, how the patient talked, or how the patient's answers made the student feel.

Engel's message cut both ways: the temptation to reduce the patient to a biological substrate was mirrored by the physician's temptation to reduce himself—to a single specialty or point of view. To something less than he actually was. In Engel's view, the physician should (as had been said of his first mentor, Weiss) "be interested . . . in human beings generally."

In 1977 Engel condensed his decades as a mediator between psychiatry and medicine into a rousing call for "a new medical model," a "biopsychosocial" model that took into account "all the factors contributing to both illness and patienthood." At the very moment when most psychiatrists were arguing that psychiatry needed to become more like medicine, Engel was proposing the exact opposite: that medicine needed to become more like psychiatry.

The question was: On what was this rousing call to be based? Research had never been Engel's strong suit, for he could not bring himself to believe in the generic study subjects on which most research was built. For him, each patient was unique. Like Gray, Engel was a clinician—a clinician in search of that eternally elusive jewel: the objective justification for subjective truth. For Engel, it was a truth derived not from clinical trials but from his own vast experience as a clinician. And by all reports he was a damn good clinician—just like his uncle Manny. A colleague once described his diagnostic skills as "almost mythical." Patients who stumped other doctors were to him as transparent as glass. For Engel had access to a sea of data that they did not—all of it highly nuanced, hopelessly individualized, and impervious to aggregation. Engel, in fact, had written an entire book on how to interview patients. The doctor, he wrote, must remain "attentive to the patient's spontaneous associations, even when at first they may seem to be irrelevant." Doctoring wasn't all action. Listening was also important. It was the same thing Theron Randolph had said years earlier. And for which he was excommunicated.

The problem was that you could read Engel's book about interviewing patients cover to cover and still not learn what Engel learned in that one moment watching John Romano pull up a chair. Michael Polanyi (who was deeply influenced by Merleau-Ponty) had argued much the same thing. The reason was that some knowledge—skills knowledge, usually, like the knowledge of the riverboat pilot—is embodied. The only way it could be transferred from one person to the next was the same way it was practiced: one-on-one. Master to apprentice.

This worked fine in the medical school where Engel controlled the curriculum, but for an academic manifesto another solution was needed. Something much more rigorous than phenomenology, certainly. As one biographer has noted, Engel was "messianic" about grounding his biopsychosocial model (BPS) in hard science, and even expressed regret that his landmark 1977 paper did not include "science" in the title. As far as Engel was concerned, if it could not be called "science" then it could only be called "art," and art (like compassion, altruism, or for that matter Heidegger's views on patient care) could only be encouraged by "inspiration or exhortation."

It wasn't enough. As any parent at the dinner table knows, once you find yourself exhorting, the game is already lost.

Engel eventually found his answer with "General Systems Theory." GST was developed by Ludwig von Bertalanffy, a philosophically inclined Austrian biologist, in the early part of the twentieth century. Like Engel, Bertalanffy grew up with a broad exposure to the humanities. As a youth, he studied Homer, Plato, and Ovid in the original, wrote plays, poetry, even a novel, and penned essays on Spengler, Goethe, and Hieronymus Bosch. It was this broad, humanistic background that (as with Engel—and Polanyi) later enabled Bertalanffy's unique contributions. (Aldous Huxley, homie of Bert, once described him as "one of those strategically placed thinkers whose knowledge permits them to strike at the joints between the various academic disciplines and so to penetrate to the quick of the living reality in a way which the specialist can never do.")

Bertalanffy first took exception to the reductive bent of science in university. At the time, in the 1920s in Vienna, the dominant philosophical mode was logical positivism, which aimed to do away with metaphysics and firmly limit truth statements to those that could be empirically verified. It was a wonderfully clearheaded sort of philosophy, or seemed to be, anyway, grandly sweeping aside all the usual nonsense that made figuring things out so damned difficult. But, coming from biology, where no part of an organism could be studied without reference to the whole, Bertalanffy knew it wouldn't fly.

While still a student, in 1924, he met a recent graduate of the doctoral program at the University of Vienna, Paul Weiss (no relation to Soma). Weiss shared Bertalanffy's displeasure with logical positivism and backed his displeasure with a solid grounding in experimental science. His key insight, from which Engel's biopsychosocial model ultimately derived, had nothing to do with humans. Far more relevant, it turned out, was the behavior of butterflies.

Vanessa butterflies, to be exact. For his doctoral thesis Weiss had studied their resting posture in response to changing gravity and light. The changing light was effected by moving light bulbs around; the changing gravity—or the effect thereof—by altering the angle of a wall. What Weiss was able to show was that the butterflies responded to both inputs—as well as the memory of previous inputs. The behavior of the insect came not as the mechanical response of one aspect of the organism to individual inputs but rather as a gestalt response of the entire organism to a range of inputs. As Weiss put it, "The elementary steps in behavior are subordinated to the state of the whole."

It may seem straightforward enough, but it flew in the face of accepted thinking in biology at the time, which, being implicitly premised on dissection, wishfully treated the organism like a corpse. Weiss, in other words, was making the same point as the phenomenologists, but with butterflies. About which it was harder to kibitz. What's more, the same implications pertained, as Weiss later made crystal clear. "It is an urgent task for the future," he wrote, "to raise man's sights, his thinking and his acting, from his preoccupation with segregated things, phenomena and processes to greater familiarity and concern with their natural connectedness, to the 'total context.'"

From there, Bert took the ball. By the early 1930s he was calling for a new approach to biology that acknowledged the organizational complexity of organisms and their embeddedness in their environments. By the late 1930s his views had developed into what he called "General System Theory" (the *s* was added later), which commenced with a recognition that the principles Bertalanffy had noted in biology could equally apply in any field in which "systems" were found.

GST came as a clear check to the reductive tendency of science and by extension to the growing trend toward specialization. "Classical science in its diverse disciplines," Bertalanffy wrote, "be it chemistry, biology, psychology or the social sciences, tried to isolate the elements of the observed universe—chemical compounds and enzymes, cells, elementary sensations, freely competing individuals, as the case may be—expecting that by putting them together again, conceptually or experimentally, the whole or system—cell, mind, society—would result and be intelligible. Now we have learned that for an understanding not only the elements but their interrelations as well are required."

Fundamental to Bertalanffy's thinking was the idea that the whole is greater than the sum of its parts. By the same token, the parts acquired meaning by virtue of their participation in the whole. In biology, this meant the organism. While most of his contemporaries were trying to account for the organism mechanistically, or else by reference to some sort of goofy abstraction like Beard's "nerve-force," Bertalanffy theorized that it emerged as a function of its organization as a system. It was this that distinguished the organism from the machine, for where machines, in keeping with the second law of thermodynamics, tended to fall apart, organisms, by drawing on external resources and building internal systemic complexity, tended to keep it together.

As Bert pictured it, this organismic complexity was tiered in structure, with the higher orders emerging from the lower ones. At the bottom you had complex proteins; then came cells, organs, and finally the whole organism. The overall nature of organisms might vary, but as systems they were all basically the same, and could all be said to follow certain rules. One example would be the more or less consistent ratio in animals between their metabolic rate and their surface area. Or between their mass and the size of their organs. And the same held true of systems in general, whether you were talking about embryological development or population fluctuations.

The tiered theme was later picked up by Polanyi, who as a philosopher was more engaged by its implications for the construction of meaning. Consider, for instance, speaking. Here the lowest tier, Polanyi said, was voice production. Above that came vocabulary and phonetics. Then grammar and syntax. And finally content, or meaning. According to Polanyi, each level provided the "boundary conditions" for the one below it. Vocabulary and phonetics, for instance, limited the scope of how the voice was used just as grammar and syntax limited the scope of vocabulary and phonetics. In this way the parts—as with Bertalanffy's organism—could be said to derive meaning from the whole. Likewise, the whole was greater than the sum of its parts—not due to some principle of organismic organization, however, but rather as a function of the tiered structure of meaning itself. This elucidation of the importance of the whole made Polanyi's argument as close to a refutation of reductive thinking as you could hope for.

Whether Engel was aware of Polanyi is not known, but he clearly intuited the same implications, which for him amounted to the long-sought path that linked mind with body—except the path, it turned out, was actually more like a ladder. At the bottom of the ladder were the body's 37 trillion cells and at the top was consciousness, identity. Only Engel took the ladder several rungs farther in either direction. Just as individual people emerged from cells, he argued, so families emerged from individual people, communities emerged from families, and societies emerged from communities. Engel dubbed this "the Hierarchy of Natural Systems," with "subatomic particles" at the bottom, "biosphere" at the top, and "person" in the middle. And insofar as any lower tier acquired meaning by its participation in and contribution to the one above it, it followed that the tiers above a person ("family," "community," "culture") were just as pertinent to the health and well-being of that person as the tiers below (nervous system, cells, atoms, etc.).

What's more, because (as Bertalanffy had argued) all systems were basically the same, suddenly it no longer mattered that humans were infuriatingly individual and complex. Nor did it matter that Engel had no data. For this wasn't a human issue—it was a systems issue. Man or bug, the same concepts applied. All derived from the same emergent structure.

For Engel, the implications were tremendous. After all, this wasn't just him opining. This was *science*. The physician, therefore, Engel would argue, with poignant militance, "has no alternative but to behave in a humane and empathic manner."

It was a bold proposal, published in a leading scientific journal, and with psychoanalysis on the decline and biomedicine ascendant it found a sympathetic audience, being cited over five hundred times in the next ten years alone.

And yet forty years later BPS survived, like phenomenology, mainly in those fields least threatened by the biomedical model, like social work, where the gender ratio was nearly as lopsided as in nursing. This was mainly because, despite the imprimatur of General Systems Theory, BPS still wasn't nearly as scientific as Engel liked to believe. GST itself, it was said, wasn't science so much as a way to think about doing science. And a similar distinction held true for BPS. It provided a way to think about the biological, psychological, and social components of health but it couldn't tell you how they all fit together—or how or when to intervene. According to one critic, it gave "permission to do everything, but no specific guidance to do anything." It was like "a list of ingredients, as opposed to a recipe."

Recipes were what was wanted. And with the host of breakthroughs in molecular genetics, brain imaging, and psychopharmacology in the 1980s and 1990s, mixing and measuring certain ingredients (the ones found among the lower tiers of the hierarchy of natural systems, like tissues and cells) became a whole lot easier. Meanwhile those professions tasked with handling the upper tiers (psychology, for instance) had made no progress at all—or nothing of the same order, anyway. The result was a widening technical and cultural gulf between high-tier professions and low ones, making it even more difficult to imagine how Engel's vision of a hybrid physician combining the insight and empathy of psychology with the potency and rigor of biomedicine might be achieved.

How could anyone possibly hope to unite a doctor of mind and a doctor of body in the same person when they relied upon such different tools? Different languages? What, for instance, did neurobiologists know about consciousness, meaning, or free will? What did philosophers know about synapses? As one critic observed, "Even the most materialist philosophers of mind do not yet ask what it feels like to be a serotonin molecule, what the serotonin molecule's reasons are for crossing the synapse, and whether or not the molecule would consider making a completely different decision."

It was like Heidegger's point about the farewell molecule, except in reverse. These languages were simply incomprehensible to each other. And the laddered path connecting them was such a long one that by the time you made your way from one end to the other you'd have lost track of where you started.

But the problem went deeper than the limits of human attention. For ultimately it came back to the fundamental nature of the expertise in which the languages were rooted. For expertise was itself just a way of rooting knowledge in identity. In other words, professional identity—the identity of a doctor, for instance—was merely a container for knowledge. Knowledge—even the fancy stuff, like mesons and bosons and quasars and blazars (a real thing)—was not just embodied; it was enselfed. Meaning that the scope of expertise was limited by the boundary condition of the self just as the scope of grammar and syntax was limited by the boundary condition of meaning. Which would mean nothing were it not for the fact that to remain coherent to itself the self required a single, coherent outlook. The self, in other words, was not built to accommodate multiple incompatible varieties of expertise. What we were capable of knowing was limited by who we were capable of being.

Even in his early days at Rochester, Engel could not avoid this truth. "In the end the student identifies with either the psychiatrist or the internist," he admitted, "and only rarely with the still abstract symbol of the comprehensive physician."

And forty years later, here we were. With the comprehensive physician no less abstract than he was in Engel's day. And (advances in integrative and functional medicine notwithstanding) evidence-based medicine still ascendant, despite the fact that "objective" assessments of health and well-being have been shown to be poor predictors of "subjective" health and well-being. And traditional organ-based medical specialties more entrenched than ever, and as unable as ever to help 82 percent of patients whose symptoms defy the dominant biomedical model. And those same patients given eleven seconds on average to explain the reasons for their visit before being interrupted by their doctors, whose burnout rates, by the way, recently increased 9 percentage points within three years, and whose suicide rates were more than double that of the general population. And many of those interrupted patients abandoning mainstream medicine altogether and turning instead to self-care or Reiki or Gwyneth. And many others like James completely untethered and roaming the land like ghosts. Or like Brian so thoroughly lost to the world that they were no more likely to be found than the farewell particle.

Here we were. In a place leached of humanism, bleak in a way that has become dully familiar, and probably by this point well beyond anyone's control. And if you dwelled on it for any length of time, it was bound to become super-depressing. But extended dwelling was at this moment not in the cards. Because Brian, it appeared, was alive.

DAY

5

CHAPTER 21
Alone with Orcas

It was like being inside a giant cannoli. I lay there trustingly in my boxers, skin glistening with 3 percent hydrogen peroxide and four pounds of pressure thumbing my eyeballs. A machine in the corner pumped concentrated O_2 as the opening bars of "Major Tom" drifted through my head. The Vitaeris 320 did not require a static bracelet and the chance of exploding was slim, but there were other dangers. J, enunciating clearly through the plastic porthole, warned me of one of them. "Don't fart in there," he said.

It was my first time inside a hyperbaric chamber. J had purchased this one for $19,000 and installed it in one of the three spare bedrooms of his million-dollar Sedona house. Was it doing anything? Did I feel better? Maybe a little.

We had arrived sometime after midnight, having drilled straight through the cotton fields after finally hearing from Brian. He'd contacted us via text as we were rolling into Tucson. A ding from J's phone and—poof—there he was, like some astronaut thought lost who suddenly began transmitting again.

J didn't seem particularly relieved to have heard from him. "Oh, Brian texted me back," he said. As if his mortal status had never been in question. And for James perhaps it wasn't. In the conditional world he lived in, people were always fading in and out. Reality was what presented through the windshield.

My own reaction—at first, anyway—was a kind of guilty disappointment. It would have been so much simpler to not have to meet Brian. To be able to conclude on a dire, angry note. *You see?* To point to his death as a symbol of something and never have to encounter the martyr himself. The failure to find him would itself be the success. Because death was a certainty, death was a measure, the greatest biomarker of them all. A death would have meant the actual end of something. It would have allowed for a moment of resonant, church organ mourning. Because for everything we had left behind—a world free of chemicals, a certain caliber of certainty, a functional government—there was never any mourning, no stillness nor farewell. Just the same panicked scramble into the future, as if the future were a tippy boat into which we could pull ourselves to escape the disturbance of being alive.

Maybe that's why I detoured through Snowflake and put off finding Brian for an extra two days. In the hope that his silence would somehow devour him and spit out the bones. But this was the thing about EI. There was no end to it. As David Reeves had said, it just kept going.

In the texts that followed we learned the full story. The problem, of course, was hell toxin. Brian had been warned about electronics from China, but Chinese cell phone batteries were cheap and he couldn't afford anything else.

Then the battery arrived and within thirty minutes it felt like he was being attacked by demons. In the two and a half weeks since, he'd been through four phones trying to rid himself of the contamination and had to bag up more than half his stuff.

We were welcome to visit, he said. It had been over a year since his last human visitor. I couldn't help but wonder what had happened to him in that time, and what this demon-addled stranger would be like in person. Would he even be relatable? If he had bid adieu to society as a whole was there any reason to think that he hadn't also bid adieu to the civil customs that went with it? Would he have gone feral, devolved into a woodland creature, smeared with feces and dirt? Why did he even accede to our visit when he had gone to such lengths to isolate himself? The Snowflakers may have been hostile at first, but them at least I could understand.

———

Now J stretched out on the floor before a flat-screen TV. His hair was wild and he wore only shorts, revealing his lean, muscular physique. He'd slept well, he said. The trip through the cotton fields seemed to have had no lasting detrimental effect. Of course we would never know why.

It was interesting to see the inside of James's house. It was one of several that he kept at various points along his route. And this wasn't counting the tiny house on wheels he had ordered or the houses in Aspen and Vegas he was planning to buy. None of them made J any less homeless. On the contrary, the more houses he owned the more homeless he seemed to become.

Still, the air was clean (the whole house hummed with the exertions of multiple air purifiers), and the deck offered a glorious view of Wilson Mountain and Steamboat Rock. J plugged his laptop into the flat-screen and a moment later Wim Hof blinked into view, looking gruff and avuncular with thinning hair and a full beard. He was sitting on a log in a forest with the sleeves of his plaid button-down rolled up. Beside him sat a handsome dog looking across the camera, wet nose to the wind.

The exercise began with deep breathing, hyperventilating the body. Then we switched to push-ups, holding our breath.

"Feel good," Hof said. "Feel strong. Make a good exercise. Go deep. You are able to do it because you have a lot of oxygen. And you are not breathing. How about that."

It felt very masculine. Man, dog, forest, beard. "Tough it out!" The strengths of Hof's world were natural strengths. But the threats were natural as well, and this was part of the appeal. You would not expect cancer to make an appearance in his videos, but you might expect a bear. Or a mountain lion. It was easy to imagine Hof squaring off against a mountain lion with relish. Even in his underwear.

The exercises were usually followed by cold showers but we decided to skip it this morning. For James, who rarely sweat, a shower wasn't needed. I'd seen him emerge that morning from twenty minutes in a 130-degree sauna looking no worse than when he went in.

Ten minutes later we were back in the Rover, en route to Whole Foods to stock up on 5-hours and lamb bars. From there we wound north out of Sedona and into the pine-thronged Oak Creek Canyon. Flagstaff lay an hour to the north and we were glad to find the wind more in our favor when we got there. The local Target was our first stop. We needed to buy a hat for Brian, and a three-quart cooking pot.

J went after the hat and pot while I prowled for Dramamine—the one drug I couldn't find in J's apothecary. It seemed I'd been living from one pill to the next since the beginning of the trip—not to mention all the 5-hours. It was another way to absorb his perspective, I guess—by absorbing all the drugs that went with it. The idea was a vaguely Engeloid one: in order to really properly diagnose someone you have to become who they are.

Maybe that explained my reaction to Corina, the young woman at the checkout counter. She was, I first felt, more beautiful than we deserved. It was an unadorned beauty but all the more affecting because of that. Her hands were slender, sliding gracefully to and fro in the gesture magicians use to make things disappear. She wore a black spaghetti strap blouse beneath a red cardigan, her blond hair up in a loose bun, arcing traces of it echoing the fine lines of her ears. Worst of all, a weighty walkie-talkie hung from the cardigan's lapel, pulling it off her left shoulder and exposing a collarbone against which there could be no defense. We stood before her awkward and stifled but she was still able to smile at us and be friendly because of the counter between us and the assurance that we would soon be gone. Which we soon were.

James seemed to have an uncomfortable relationship with women. He desired them (one morning I was startled to see a woman's bare posterior pop up on the Rover's navigation tablet), but there was something almost cartoonish about this desire. It was as if women were another species to him, with their own mysterious language, habits, habitats. There were none, for instance, to be found in Sedona, he said, whereas L.A. was brimming with them—although the strip clubs there, he'd said, were no good.

This was mildly shocking, to learn that he patronized strip clubs. He'd already mentioned that he used to hang out at strip clubs when he was dealing cars in Atlanta. At that time—he was nineteen—the height of glamour was to pull up in front of a high-end club in a really sharp car. He had also (in answer to a query about his ability to be firm in business) shared a story about turning away a short brunette stripper at the door of a friend's bachelor party after he, James, had explicitly ordered a tall blond one. As if the waitress had made a mistake with his entree.

Possibly his understanding of women derived from the fact that he had never really seen what a healthy relationship looked like—that is, if the stories he'd told about his father could be believed. In one, for instance, his father sent him to the local airport to pick up two girls he had flown in on a charter flight from South America. His father often traveled to South America, James said, and dated a lot of women there. After the divorce he married a thirty-five-year-old woman from Peru. Then he divorced her and married a twenty-five-year-old Filipino woman. At one point he was dating a nineteen-year-old. One summer, when James was fourteen, they all lived on a boat together—James, his brother, his dad, and the nineteen-year-old.

To James, all of this seemed perfectly natural. When I asked him about it, it wasn't the age of the women he objected to but rather their origin.

"I didn't know what to think," he said. "Sometimes I thought, 'Why couldn't he just find a girl here?'"

James had planned to get married himself at one point. This was in Aspen in 2004, when he was twenty-nine. I was surprised by this and asked who the woman was and how he had met her.

"I think she was attracted to me because . . . she said something to me and I was just kinda, like, yeah, whatever," he said. He gave a derisive little chuff, but it wasn't clear who the derision was directed at, the woman or himself.

"What really happened is, I drove to the grocery store in my brand-new Bentley," he said, chuffing again. "And I had on, like, an Under Armour shirt, y'know, and I was in great shape. And she was with a friend of hers. And she saw me get out of the car and she's, like, 'Nice car.' And she had a Mercedes convertible and I was, like, 'You too.'" The derisive laugh. "'You too.' Kept going. I didn't, like, stop and talk to her. But she was beautiful. And her friend was hot, too. And I was, like, 'I can't believe I'm such a dumbass!' Like, like, I wish I could talk better to girls. I can't do that, y'know? Why am I such a social misfit?"

At which point it was no longer unclear who he derided.

James saw the woman again the next night at the Caribou Club and they got to talking. She was a model, or former model. She wasn't doing anything in particular in Aspen. She was just trying to get through a separation or divorce. I could imagine all this clearly enough. Curiously, though, when I tried to summon up a question or two to help enhance the picture all I got was a blank. It's like there was nothing there to latch on to. The blankness of the story was overwhelming, and I couldn't find a way past it.

"Why did it end?" I finally managed to ask.

"Um . . . I don't know," J said, after a long pause. "She wanted it. And . . . I was shocked."

"She surprised you? She just said 'I want out' one morning?"

Struggling for words, he probed the opacity, like a half-remembered dream.

"She said she felt like . . . she had missed out on her twenties."

This after they had been dating for two years, and had talked about getting married. It was bizarre. In the story I was picturing they didn't even seem like real people. More like cutouts that someone had suddenly turned to face different ways.

The blankness, the inarticulateness, the self-hatred and inability to connect lined up almost too well with van der Kolk's version of trauma.

"You don't need a history of trauma to feel self-conscious and even panicked at a party with strangers," van der Kolk wrote, "but trauma can turn the whole world into a gathering of aliens."

It was the same sort of disconnect I'd felt from James in Snowflake.

Earlier that morning, before leaving Sedona, he had admitted that he'd never been particularly social—although he did like to be around other people. By way of example he cited orcas.

Orcas?

Yes. He'd recently seen a documentary about orcas while lying in a hyperbaric chamber back in Aspen. The orcas all whaled along together. Obviously they enjoyed each other's company because they all belonged to a pod.

The documentary had been recommended to him by the medical director of the facility that housed the chamber. And on hearing the story I was struck by a vision of James, lying alone in his pressurized chamber, still quietly kicking himself for the bungled exchange with the beautiful medical director as the orcas arced companionably through the waves.

Brian, too, had difficulty relating to women. He had been married for twelve years before the EI hit, but from what he'd told me the marriage had always been hollow. Like James, he, too, had a hard time connecting—and not just with his wife. Unlike James, however, he was pretty good at faking it—up to a point.

"I know looking back that as soon as it got to that level of 'Hey let's hang out,' whatever that level of connection was, it scared the shit out of me," Brian had said in one of our phone conversations. "I was overwhelmed by it."

At work, for instance, he had three friends—Chad, Doug, Chris. But he only very rarely hung out with them outside of work. The two times he did, he drank heavily to hide his unease.

"Like I always did when I hung out with anybody. Because that was the only way I could fake a connection."

But this is where Brian's and James's stories parted ways—at least if you bought the PTSD argument. For if James's EI was an expression of his PTSD, for Brian it helped bring it to a head. For it wasn't until he developed EI that making a real connection—and actually being heard—became important. By that point, however, no matter how loud he yelled no one seemed to hear. His words, he discovered, meant nothing. His in-laws thought he was crazy and wanted him put in a psych ward.

The breaking point came on Christmas Day. Brian was in the backseat of the car, driving with his wife and mother to his in-laws' house. He was trying to explain why he needed to stay there, because it was the one place he could tolerate.

"I was trying to tell them, and they just kept talking about other things. I was in the back lying down, and I remember I took my fist and I just started punching the door. And I said: '*Nobody. Is fucking. Listening. To me.*' And blood was flying everywhere because my knuckles, from hitting the door. My mom, I remember, started crying. She realized I was crying out for help. And my wife was, like, 'Brian gets like this when he's just not feeling good, he'll throw a tantrum or whatever.'"

She ended the marriage soon after that.

"'We're done,' she said. I'm, like, 'We're done? We've known each other for eighteen years. We're just done?' And she said, 'We've never had a relationship.'"

At that point, unable to work, rejected by the medical world, rejected by his own wife, Brian's aloneness was complete. It was what you might call a perfect aloneness. Like the aloneness of James watching orcas in his chamber. There is a dewdrop fragility to such moments of aloneness. Homeless at this point, Brian described walking one day to the shore of Muskegon Lake. There he met another homeless man who was sitting on the rocks, fishing. They began talking, and the man, who was older, African American, began sharing a few things about his life. He lived in the corner of a warehouse. Social Services was trying to help him but he liked it where he was. He had a pet raccoon.

The lake was still and for once Brian was, too, open like a bell to receive the toll.

"After I left, I felt I was concerned about him. And that generally wasn't something that I ever felt. It was like this natural bond that I'd never felt before. And it felt right . . . I remember I was thinking, 'What can I give this guy. I want to buy this guy something.' I had never . . . I wasn't that thoughtful. I always struggled to remember anniversaries, my wife's birthday, I was always kinda like, 'Ah shit, gotta go get something, gotta do the obligation.' Where, at this point I was, like, 'Man, I love this person.' I know it sounds weird but I never experienced that, genuinely, in my entire life."

He never saw the man again. But something in him had finally shifted. And what his wife had always been saying finally began to register.

"She would make comments throughout our marriage that she would never have gotten married if she knew what I was like. And I kinda didn't know what she meant. But I didn't know what a relationship was. It was like I was a five-year-old. It was all about me before then because I didn't know how to connect in that way, y'know, out of myself."

It was a familiar and satisfying story, a story of adversity from which a man emerges stronger and wiser than before. In this sense, Brian seemed farther along the road than James, who hadn't yet found his way out of the chamber. Though less affected by EI than Brian, James in this sense seemed the more afflicted. For in the silence that had come after everything was taken from him Brian had attained insight and made a kind of peace. He had apologized to his wife and was hopeful she could move on with her life and find a way to be happy.

"Sometimes at night when things are dark I still find myself saying I'm sorry to her, over and over again," he told me. "Because I really am, and I wish I didn't fuck her life up, because those were her childbearing years, and she really wanted to have kids."

And all of this would have been very satisfying indeed but for one thing. Which was that in Brian's telling of the story the cause of his transformation was actually not his Jobean suffering, per se, or the emotional insight that came with it. The cause, he said, was that he had finally gotten clear of all the toxic chemicals and mold.

"It wasn't like my wife left me, and *It's a Wonderful Life* started happening," he said. "It literally was the physical changes in my brain. I could just tell."

Not only this, but in his view the reason his wife didn't accept this explanation was not because she thought it was nuts but because she had already given him a hundred chances and had no more to give. Brian's certainty was such that the idea that his wife might doubt his story line never entered his mind. And so what seemed like a rich story of emotional growth was undermined by the suggestion that Brian's early personal shortcomings were not his fault. No more than head pain was the fault of someone suffering a migraine.

It seemed like the same sort of mistake so often made by EI skeptics, when the need for certainty ended up limiting conclusions to the either/ or variety—medical or psychological, body or mind—when the reality was far more complicated and multifactorial. It made you wonder whether it was this same need for certainty that enabled EI to develop in the first place—and prevented it from being cured.

CHAPTER 22

The King of the Forest

Beyond Flagstaff the hills settled as we roved into the empty flatlands. The sky was hard blue but cushioned by thin clouds at the horizon. The grass was baked brown but green along the roadside where the runoff allowed a little color. Small mercies. We drove, the dashed lines and telephone poles keeping tempo through the wastes.

Brian had texted his location—somewhere near mile marker 576 in the Kaibab Forest, a few miles down the road from Jacob Lake. He had already told me a little about the place. He'd first learned of it from a doctor friend in similar straits. The doctor belonged to that xenophobic class of sensitives who didn't trust anyone who hadn't experienced EI firsthand. I'll call him Dr. X. Brian had met X through the EI Facebook groups, where the doctor would sometimes chime in to offer advice when Brian was feeling suicidal.

The doctor was an unusual character in the EI community in that, as well as being among the most severely affected (at one point he was compelled to abandon his wife and two small children for eight months to search for a place to get well), he was also endowed with all the credibility and expertise conferred by a 2012 degree from Tufts Medical School. The same uneasy admixture of perspectives could be found in Dr. Hope—but Hope had recovered, and found a way to anchor her expertise in an established practice, with an office and appointment calendar and all the rest. Dr. X, meanwhile, was still practicing out of the trunk of his car and experimenting on himself like Jekyll (a brilliant physician, let us not forget).

Dr. X had been fleeing south from Zion where the winds had stirred something up when he noticed a change in the air, as if someone had just thrown open a window. Perhaps it was the filtering effect of the pines, or the eight-thousand-foot elevation, but there was something different about this place. Dr. X wanted to study it, prove that it was different. Like James, what he wanted was data.

When Brian arrived he immediately got sick. This was to be expected. Paradoxically, pristine environments often had this effect. Herxing, they called it, a reference to the Jarisch-Herxheimer reaction, which usually referred to a side effect of antibiotics in which the bacterial wreckage of destroyed microbes filtered into the bloodstream and triggered an immune response. It got so bad he couldn't even stand to be near Dr. X. They camped a few hundred yards apart in the forest, donning trash bags and breathing masks and hollering at each other from a safe distance when they needed to talk. By this point the forest had come to resemble an impromptu triage, with medical equipment everywhere and IVs hanging from the tree limbs.

When the forest rangers got wind of the encampment they were not pleased. They arrived one day when Dr. X was elsewhere. Brian was in the middle of trying to explain when Dr. X pulled up and the ranger, already spooked by Brian's Gandhi vibe, drew his gun. It was a surreal moment but certainly not the first for Brian and Dr. X. They subsided into the stillness of the forest until the ranger had a chance to calm down. When he finally did, he gave X twenty-four hours to gather his gear and clear out.

As for Brian, with the ranger's permission, he stayed on. The place worked for him. Besides, as he told me later, "I'm sick of running. I can't just run from everything in my entire life."

———

Our path now took us through the bleak lands east of the Grand Canyon. Great salmon-colored bluffs began edging in from the horizon and soon they were rearing over us on either side. At around two we crossed the Colorado River and began the winding climb onto the Kaibab Plateau. Here the road was faded like old denim, and molten veins of crack filler stood out sharply against the gray. As we ascended, scrub pine began dotting the hillsides. At mile marker 567, a sign warned of "moderate" fire danger, and to the west I briefly mistook an elastic flock of starlings for smoke.

At last it arrived—mile marker 576. A dirt road led into the pines. Their shadows flickered over us as we steered off the asphalt. The pines were tall ponderosas, dark green and amply spaced. I lowered the window and sucked a few lungfuls of the fresh forest air. Dr. X was right, the air did feel different—dry, clean, washed by light. "The holy grail of air," Brian had called it. The kind that one could breathe, as Willa Cather once said, "only at the bright edges of the world." There was nowhere else to go from here. With his back to the Grand Canyon Brian had withdrawn to the limit.

As the last trappings of civilization faded behind us I felt ambushed by an odd sense of vulnerability. It was a disconcerting feeling. There is a certain kind of vulnerability that one forgets is possible. This is the social vulnerability of adolescence, which you can still sometimes feel in those rare moments when you are so thoroughly displaced from familiar territory that none of your social capital obtains, and you suddenly find yourself unsure of who you're supposed to be. At such times there is little to do but hold the course and trust to circumstance.

We'd been nosing along the forest trails for twenty minutes and I was beginning to think that Brian might not even exist. Like maybe he was just another recording, like the voice from the meteor crater near Winslow. But then the trees parted and there he stood, skinny as the Nazarene.

James parked the Rover and we stepped out. Up close he looked dangerously thin, almost elongated, as if from having spent too much time among trees. His face—red with EI sunburn—was also long, but rounded looking, swollen like a hungover drunk's. His eyes were small and puffy and his lower lip protruded wetly. The veins on his arms stood out.

He greeted us with a half smile and precarious eye contact, the way you greet those with whom you share a lot in common and yet have never previously met (fellow friends of a dead man, a new college roommate). I followed him through the trees, taking in the immaculate whiteness of his tee shirt, the blue nylon basketball shorts, the bath slippers scuffing the pine needle carpet.

He had two tents, a larger, room-sized one and a pup tent. Beside the larger one a half dozen kitchen garbage bags were heaped like an offering around a three-foot-diameter tree. These, Brian said, contained the clothes and other possessions that had been contaminated by hell toxin and had to be thrown out. The only thing he wasn't throwing out were his two coolers because they had all his food in them, but now as a consequence he suffered whenever he went near his cooking stuff.

"All my silverware, all my dishes. The reason we call it hell toxin is that you can't wash it off with hot water and ammonia. It just grows. It's like gremlins. So, I'm starting over. I have eight white tee shirts and this is my last pair of shorts."

He had set up a couple of folding chairs, one of which he had just bought so he could be sure it had no hell toxin on it. I settled in one of them as James excused himself to get a snack from the Rover.

"Absolutely," Brian said. "Eat or take whatever supplements you need, that's all I do all day long. So I understand."

"Yeah, I just got high levels of ochratoxin from a place I rented—" James said.

"Aw, man."

"—so I'm, like, trying to do the detox."

"I hear you, man," Brian said. "I'm sorry about that."

It was interesting to watch them interact with each other, the dance of their finely wrought courtesies—James's Southern decency and Brian's spacey Midwestern kindness.

J returned a minute later and presented Brian with the three-quart pot (we couldn't find a hat). He had begun to start work on an early supper—boiled cubes of butternut squash with garlic salt. The forest was quiet around us. No wind, just the high, holy hush of air combing through numberless pine needles. The sun warmed the rough bark of the trees and rearranged masks of light on the forest floor, light and silence opening to accommodate a billion thistly details. Reality peacocking, showcasing the full extent of its powers. A horsefly ornamented the air with black necklaces.

"I love the evenings and I love the early mornings," Brian was saying. "I love the quietness and stillness. I had a chipmunk this far away from my sandal. He was nibbling on the rind I left from my honeydew melon."

Chopping squash on the lid of a giant Tupperware bin, Brian seemed almost jarringly informal, a priest washing dishes in the vestry. Indeed, what the Kaibab felt most like was a cathedral, in which the one unspoken word was "behold." Behold the pinecones individually placed. Behold the noble trees structuring the light into columns. Behold the buxom silence and the presence of God.

Much of what Brian did here, he explained, was pray. Not that he had come here looking for God. He simply appeared one day, doe-like on the fringes, after the brain retraining had silenced enough of the noise.

"I'm just, like, wow, I totally feel Him," Brian said. "I'm not just crying out to Him because I'm miserable. I'm sensing Him in everything."

And despite the teenage phrasing, in the roomy quiet of the pines his words took on a certain power. The power that comes when the words arise from the realest part of you and enter the world intact, unweighted by doubt or secondary intentions, bearing nothing but the simple meaning with which they were dispatched.

"I get a different perspective of Him here," Brian went on, "more mature, not just that He's somebody I cry to when I feel like I'm gonna die but that He's just this peaceful constant that created me because He wants to hang out with me and doesn't really want to ask for anything else other than that."

Religion wasn't new to Brian. When he was young his mother taught Sunday school and made him memorize Bible verse. It was confusing at the time because the Christian message he was getting at church did not exactly jibe with the message he was getting from his father at home. Religion didn't really begin to mean anything to him until college, when he underwent a conversion experience. But that, too, faded into the background, and it wasn't until he was sitting on the shore of Muskegon Lake, feeling love for a random stranger and his pet raccoon, that he found himself thinking about God again. Now, thirty-odd years after Sunday school, Brian was reciting phrases from the Bible once more.

I had never much cottoned to religion. It was one thing to acknowledge the terrific volume of all that remained unknown in the universe, but to gather it all under the robes of a single deity seemed a bit wishful. Far more likely this volume was disorganized and broadly dispersed. Bemused agnosticism, it seemed to me, was the only sensible view.

Besides, there was something ridiculous about the way God was, like the government, always turning up to underwrite outsized liabilities, whether Descartes' Third Meditation or Brian's will to live.

Yet something about the way Brian talked about God made me feel that these objections were all beside the point. In the russet stillness of the forest what came into my head instead was an image of a girl placing her hand on the neck of a horse. The image bubbled up from a story van der Kolk had told about a fifteen-year-old girl with an extensive history of sexual abuse and multiple suicide attempts. Her medical charts described her as mute, vengeful, reckless, and ultimately unreachable. There was no way to extricate this girl from the living trap she had become. Until someone had the idea of giving her a horse to look after.

The horse was a benign creature—wonderfully armless, unable to grab you, sturdy and true. The horse would stand still for you. You could groom a horse, run a brush down its silken flanks. You could talk to a horse and its ears would flick. You could cry on the neck of a horse. The horse wouldn't mind. You could approach the front of the horse and look into one of its large, mysterious eyes. It would look back. It would see and not see. The horse let the girl feel care. To feel care and to feel cared for were essentially the same. One was just a version of the other.

Why did Brian have to come so far to feel such care? This was the question. The world is dappled with loci of care. Families. But families fail. Doctors' offices. But doctors, too, fail. Patients come through the door like James coming home after his motorcycle wreck and are just as brusquely dispatched, bleeding under their clothes.

I thought of that Texas doctor accused of prescribing vodka drops. *If it would have helped her I may have used it, and I'm serious about that, I may have.* Something about the way he said this, as if he were asserting by way of reminder the nature and extent of his sovereignty, laying down a boundary which others were forbidden to cross. And with this the place contained by the boundary was called into being, a place of inviolable silence into which the patient could enter and be noticed. A place not unlike this forest.

To me, it suggested that the doctor's job was not just to defend the patient (against germs and viruses and cuts and bruises) but also to defend the sanctity of that place—from all those who did not have the patient's best interest at heart, but also from the obtrusive welter of dualities with which medicine was plagued (mind and body, psychologist and biologist, clinician and researcher) and which were ultimately just an artifact of language. In the silence they subsided. In the silence Brian was neither beauty nor crone, sane nor crazy. The seeming incompatibility said more about the limitations of language than the nature of that which it struggled to describe. It wasn't language—the ability to speak—that made the doctor but rather the opposite—the ability to listen.

While it might be true, in other words, that what we were capable of knowing was limited by who we were capable of being, there was another way to look at it. Which was that who we were capable of being was limited only by what we allowed ourselves to hear. The patient primarily, for every patient was different—but also the stories of everything else in the world that we did not already understand. The behavior of butterflies, fault tree analysis, Paiute forestry, the triad treatment, the history of graham crackers, farewell molecules. This is where the breadth of perspective came from. The greater the listening, the broader the perspective; the broader the perspective, the better the doctor could distinguish what was important from what wasn't—the dictates of her own expertise included. Did it matter if the pain wasn't "real" if it still caused suffering? Did it matter if the meds weren't "real" if they stopped the pain? ("One of the most successful physicians I have ever known," Jefferson wrote, "has assured me, that he used more bread pills, drops of colored water, and powders of hickory ashes, than of all other medicines put together.")

The good doctor brought the listening with her wherever she went. Whenever she entered a room it was as if the forest entered with her. It was this that so struck Engel when he observed John Romano pulling a chair up to a patient's bedside. In a way, that's all the road trip with James had ever been—an 1,800-mile chair drag. Here was the true genius of the doctor: she could cover that same distance in moments, when for me it took five days.

It would never be easy for doctors to devote so much energy to listening. There was already too much to know, and too little time to know it in. How could you justify hearkening to Husserl when you were already four months behind on your professional journals? How could you justify a twenty-minute intake when you had sixteen other patients to see? Listening might very well be edifying, but it was also inefficient.

But this overlooked the obvious. Which was that inefficiency was ultimately a far lesser danger than the alternative. Losing perspective. Losing control. Becoming someone else's tool, and allowing some other random factor to displace care.

Care was what knit the world together. Care was a source of conviction. Care was what kept us from being seduced by our own intelligence, by money and prestige. Care was not easy. But if this trip showed anything, it's that that silent place was still there for whoever wished to claim it. There was after all something sort of holy about the silence of the silent place. Which was really just a way of saying that this silence deserved to come before everything else.

We talked for several hours more as the sun descended through the pines and the air began to cool. High above, the wind parsed the forest

eaves, searching perhaps for one needle in particular. Despite being surrounded by trees, here, too, it felt like you could see for miles in any
direction.

To my surprise, Brian said he was thinking of leaving the Kaibab and
moving back to Michigan. It was a hard transition to fathom—from paradise to Muskegon. But he'd been out in the wind for many months and
he could no longer tolerate the solitude. Recently, he said, he drove two
hours to get a haircut. After years in the woods it was worth it just to
feel someone's hands on his head. Like the caring hands of a doctor. Like
a benediction. What he needed was the same thing James needed. The
same thing everyone needed. To connect with other people. To be loved
by other people. For, great as He was, it was not enough to be loved only
by God.

CHAPTER 23
Goodbye to the Pines

And I would be lying if I said that some part of me did not want to stay up there among the pines with God and Brian and even the mad Dr. X if he was into it. And clad myself in trash bags and meditate and howl at the moon. It was what some unruly part of me had always wanted, to cast off the old rules like moldering clothes and run screaming from civilization, to cast off language itself, which always promised freedom and yet so often seemed to deliver the opposite, to become a true maniac, to become like a mountain lion and roar a maniacal roar. Maybe it was something like this that Hof felt when he jumped into the frozen canal that day. Not just a desire to escape but exuberance, *joy*. All that which could not be systematized and contained. Breathe, motherfucker.

Brian agreed to a handshake and suggested we stand together and take pictures. So we did, James and I taking turns with the camera, Brian leaning in, castaway thin, three kooks inanely acting out one of civilization's cheesiest customs.

The Rover rocked as J piloted us back through the maze of dusty roads and onto the asphalt. The sun ran beside us through the trees, its long arms lifted in farewell. As we came out into the open we could see it looking small against the greatness of the land, a little orange dent in the far hills where it was setting. In moments it was gone, but as the land around us faded the sky held on to its blue glow as we began the long descent into Vegas.

In the release that came with finding Brian our talk drifted aimlessly from one subject to the next. The recent suicide of a sensitive whose toxin report showed the same levels as James's. The Adult Video News Awards which J attended this year in Vegas, where he had to change hotel rooms four times. His memory of buying Van Halen's *1984* as a kid, and how his mom cut off the cover because it showed an angel smoking a cigarette.

"I guess I was nine," James said. "My mom was a nurse."

The silence pooled. Then he surprised me by saying that while I'd been talking with Brian he'd received a text from his dad.

"I told him about my recent exposure," he said, his low murmur nearly lost in the hum of the road. "He just said he was thinking about me and hoping I'd get better. He's never said anything like that. In the old days there would be just silence, or he would have walked over me. It was almost like . . . being nice or . . . understanding was almost like . . . sissy or something."

I waited, not wanting to interrupt the flow. But there was no flow. It was the same hung moment that followed his encounter with Charlie. This was the moment James lived in, the one that stopped right before fury.

But by this point we had been through enough that I felt I could ask: whether it might help to talk to his father about this. See if he could own up to the kind of father he'd been. Perhaps even tender some kind of apology.

James surprised me again, saying that his father had apologized, once. Or tried to. But James couldn't accept it.

"He said, like, 'I'm sorry I wasn't a good dad,' and I said, 'No, no.' I was just really uncomfortable and I didn't want to . . ."

"Why?"

"I dunno, I didn't really wanna say . . . 'Yeah you were not.' Or . . ."

"Why not?"

"I don't know. I think I said something like, 'I'm sorry I wasn't a good son.' I just wanted it to be over and, like, let's move on."

"You were protecting him."

"I guess so . . . It's a weird thing but . . . I just didn't know how to handle that situation. It's a little fear or something that . . . Like I said when I didn't accept his . . . apology . . . it was this awkward, like . . . I just . . . kinda, like, really quickly, let's change the subject, talk about a deal or something, whatever I did to, like . . . Okay, I'm glad it's over; now we can talk about something else?"

"It's hard to forgive your dad," I suggested. "There's something painful about seeing your father smaller than you are when he was such a big figure, growing up."

"Yeah . . . my dad was the man. He was, like . . . the one I wanted to be like, the figure, the most influential in my younger years."

"Yeah, but it sounds like he was also a withholder."

"Yeah . . . and maybe he didn't know how to express himself to me, like I didn't know how to express myself to him."

"You're still protecting him."

"I sorta am. Because I don't really know . . . What's . . . right. What's . . . wrong in this specific instance . . . He didn't, like . . . ever . . . like . . . like physically hurt me. So that's one thing, but then . . . I dunno, was I . . . I was . . . let down a lotta times, yeah, but was I expecting something unreasonable? I dunno; maybe I was. I'm trying to figure it out. Was I expecting something more than I should have expected?"

There was something excruciating about this. It was the core, it seemed to me, of James's uncertainty. And for a moment I saw him clearly as the boy he was, and felt for him in that way—the way you feel for a boy trailing after his father as he's being yelled at. And the boy just kind of creeps along. Because what else can the boy do?

How should James know what was reasonable to expect? How should any child? Our parents set the template for how we experience care. They let us know what we deserve.

J had a very clear idea of what his body deserved. But he still wasn't sure what he deserved as a person. As a son.

DAY
6

CHAPTER 24

Bernhoft's Judgment

From Vegas it was a long straight shot to L.A. I was operating on three hours of sleep, Illy coffee, Addies, and a gnarly microwaved burrito from the local Chevron. You know you've been on the road too long when you start eating at the same place you gas up.

J wasn't looking so hot either. He was quiet as we hit the four-lane, eyes bloodshot, throat working. About an hour in, he reached back and retrieved some tubing which he clipped to his nose. Then he uncocked the oxygen tank that had been rolling around behind us since leaving the St. Regis, and the cabin was filled with the quiet hiss of air.

The emptiness we drove through now was a more familiar emptiness, a common passage used by millions fleeing one great city for another. Despite its desolation the desert here felt like a space we had already reckoned with. We knew what to do with this sort of wildness. One moment we were passing a huge solar array. The next, a giant windmill blade strapped to the back of a long truck was passing us. I gazed into the distance, stupefied, treating my Addie-induced dry-mouth with pomegranate 5-hour and wondering how to feel.

After three hours of nothingness malls began to appear. Then came the first Walmart pucker, the great sphincter of America. By eleven we'd hit the red line on the GPS and J's leg started to waggle, as it always did when he needed to pee. We watched civilization accrete around us, staring like tourists at the fructifying car dealerships, landscaped embankments, and flashing signs warning of drought. Entering Santa Monica J pointed out Bundy Drive as we passed over it, home at one time to Nicole Brown Simpson, whose murder inspired another great American road trip. By this point his leg was revving in high gear, and at the next opportunity he took us off the highway and pulled over to let loose between two parked cars. Here was something else to be said for the desert. It was far easier there to find places to pee.

We stopped at one of Dave Asprey's Bulletproof Coffee cafés for lunch, where J, to my surprise, with his fit torso, golf visor, and brainy architect glasses, blended seamlessly with the rest of the clientele, the ponytailed girls in yoga pants, the Under Armour–clad fitness dwarves, the kid with the tatted arms and black tee saying "Butter makes me happy," as if I gave a shit. Here, all the stuff that James was into suddenly seemed not only unremarkable but fashionable. Espresso "with Mentalist roast," Grass-fed sliders "served with avocado on a sweet potato zucchini unbun™." This is where we had landed, finally. In the land of the trademarked hamburger bun. Side effects, the menu warned, could include Excessive Productivity, Limitless Energy, and Mental Clarity. There was also a vibrating plate you could stand on if you were so inclined, for what purpose I could not divine. "The NASA astronauts use it," a white kid with dreads explained.

L.A. A city of people who felt fully entitled to take care of themselves. By whatever bizarre means they could imagine. No wonder James felt comfortable here. Just another freak in the freak kingdom, as Hunter Thompson would say.

"It's the largest group of misfits that I've found," J said.

———

A half an hour later we arrived for the appointment with the doctor in whom James had placed his last hopes. His office was located in the back of an unprepossessing building in downtown Santa Monica. The door opened onto a small waiting room furnished with a few seriously off-gassed-looking chairs. Behind a sliding window a receptionist warmed her hands on a mug of tea, and from somewhere we could hear the tinkle of wind chimes.

Eventually we were ushered in to see the doctor, who, following the usual pattern, used to be a pancreatic surgeon before EI laid him low, whereupon he retrained in environmental medicine. "No gynecologist should be allowed to operate until he has had a hysterectomy," Bernhoft (quoting his mom) declared on his website—and the same (or something similar) held true for EI docs.

Bernhoft's office was large, with louvered windows and pale yellow walls. A green examination table occupied the center of the room. A skeleton stood in a corner.

Bernhoft looked to be somewhere in his sixties. I can't say he projected much of a listening air. His head was pointed at the floor most of the time as he puttered around collecting ingredients for James's injection, and his talk came forth in a sarcastic murmur, which, though not at James's expense, seemed mostly designed, like Charlie's talk, to amuse himself.

Much of Bernhoft's talk concerned the nefarious behavior of various pharmaceutical companies and government agencies.

"And Phil looked under his car every day for the next year," one of these stories ended. "That's the kind of business we're in."

I had no reason to disbelieve him.

This was to be James's third injection of five. Each injection contained 450 allergens—a "one-size-fits-all approach"—and was delivered a number of weeks apart. Bernhoft promoted this method over that of Gray and Rea, which he described as "a Band-Aid."

"Bill Rea's been giving himself shots for thirty years," he said. "I did it for a year and a half and I was asymptomatic. I could go to Macy's and survive. QED."

James listened deferentially and then, laying down his arm for the injection, asked the question that had been weighing on him for the last thousand miles.

"Assuming this works, I should be back to normal without the sensitivities, right?"

"Yeah, 80 to 90 percent likelihood, yeah," Bernhoft said.

And James bowed his head.

And indeed soon reported that he was feeling better. A lot better, he said.

From Bernhoft's office, we proceeded to downtown Santa Monica, where James and I parted ways. The eddy he was caught in was sweeping him back out into the desert, and it was time I disembarked.

We managed an awkward half hug. And both proved unable to guess what the right words might be. It made me appreciate the cheesy photos we'd taken with Brian. Like a doctor's diagnosis these customs covered for us, bridged the inexpressible.

"We'll do it again," J said.

In the company of his fresh absence I walked dazedly through downtown Santa Monica. Smells assailed me. The charcoal burn from a restaurant. Patchouli from some hippie bead stand. A whiff of leather from a leather shop. Apple shampoo. Scents coming and going so fast I couldn't identify them. My body moving through the world, the world moving through my body. After living for a week with the unending calamity of EI I was aiming for the end of something. Eventually my wandering brought me to the Santa Monica Pier. I walked its length, relishing the sharp punctuation where it met the sea.

Acknowledgments

First and foremost I want to thank all the people living with EI who shared their thoughts and experiences with me. I have tried to honor their trust by doing my utmost to understand EI in all its complexity. In particular I want to thank James and Brian, who selflessly participated in this project in the hopes that doing so would prove useful to others—people living with EI, their friends and family, and the countless medical professionals who have pledged their lives to the alleviation of suffering. I also thank the folks at Snowflake: Liz, Scott, and their friend, with whom I spent several fascinating hours talking, eating sardines, and gazing across the darkening Arizona barrens. Also David Reeves, marooned by EI in the middle of nowhere and dreaming of his lost life in New York. I am grateful, too, for the insight provided by all the healers I spoke with in researching this book, especially Dr. Jeanette Hope, Dr. Michael R. Gray, and Dr. Robin Bernhoft.

This book leans heavily on the previous work of others. For the chapter on forest fires I am indebted to Timothy Egan's *The Big Burn: Teddy Roosevelt and the Fire That Saved America*. For the chapter on William Perkin I am indebted to *The Rainbow Makers*, by Anthony S. Travis, *Mauve: How One Man Launched a Color That Changed the World*, by Simon Garfield, *The American Synthetic Organic Chemicals Industry*, by

Kathryn Steen, *Dupont and the International Chemical Industry*, by Taylor Sudnik, *The Emergence of the German Dye Industry*, by John Joseph Beer, and *Madder Red*, by Robert Chenciner. For the chapter on Fritz Haber I am indebted to *Master Mind*, by Daniel Charles, *A Higher Form of Killing*, by Diana Preston, and *Fritz Haber: Chemist, Nobel Laureate, German, Jew*, by Dietrich Stoltzenberg.

The chapter on Benjamin Rush draws on *American Physicians in the Nineteenth Century: From Sects to Science*, by William G. Rothstein, *Benjamin Rush: Revolutionary Gadfly*, by David Freeman Hawke, and *Revolutionary Doctor: Benjamin Rush, 1746–1813*, by Carl Binger. Also, *Lotions, Potions, Pills, and Magic: Health Care in Early America*, by Elaine G. Breslaw, *Pseudo-Science and Society in 19th-Century America*, edited by Arthur Wrobel, *The Great American Medicine Show: Being an Illustrated History of Hucksters, Healers, Health Evangelists and Heroes from Plymouth Rock to the Present*, by David Armstrong and Elizabeth Metzger Armstrong, and "Sanguine Practices: A Historical and Historiographic Reconsideration of Heroic Therapy in the Age of Rush," a journal article by Robert B. Sullivan.

For the chapter on risk I owe thanks to Karen Zachmann for her paper "Risk in Historical Perspective: Concepts, Contexts, and Conjunctions," to Arwen P. Mohun, for her book *Risk: Negotiating Safety in American Society*, to Robert N. Proctor, for his book *Cancer Wars: How Politics Shapes What We Know and Don't Know About Cancer*, and to Soraya Boudia and Nathalie Jas, who wrote several fascinating chapters on the history of risk.

The chapter on tuberculosis and neurasthenia could not have been written without *Seeking the Cure: A History of Medicine in America*, by Ira Rutkow, *Bargaining for Life: A Social History of Tuberculosis, 1876–1938*, by Barbara Bates, *Chasing the Cure in New Mexico: Tuberculosis and the Quest for Health*, by Nancy Owen Lewis, and *Health-Seekers in the Southwest, 1817–1900*, by Billy Mac Jones. Also, *Neurasthenic Nation: America's Search for Health, Happiness, and Comfort, 1869–1920*, by David G. Schuster, and *American Nervousness, 1903: An Anecdotal History*, by Tom Lutz.

The chapter on Ludwig von Bertalanffy and George Engel draws on *John Romano and George Engel: Their Lives and Work*, by Jules Cohen and Stephanie Brown Clark, and *Uncommon Sense: The Life and Thought of Ludwig von Bertalanffy (1901–1972), Father of General Systems Theory*, by Mark Davidson.

I should also acknowledge *Another Person's Poison: A History of Food Allergy*, by Matthew Smith, *What's Gotten into Us?: Staying Healthy in a Toxic World*, by McKay Jenkins, *Toxic Safety: Flame Retardants, Chemical Controversies, and Environmental Health*, by Alissa Cordner, *Bodies in Protest: Environmental Illness and the Struggle over Medical Knowledge*, by H. Hugh Floyd and J. Stephen Kroll-Smith, and *Legally Poisoned: How the Law Puts Us at Risk from Toxicants*, by Carl F. Cranor.

Countless journal articles by a broad range of scholars have also nourished this book. I am grateful to them all.

Lastly I want to thank Laurence Thomas, a philosophy professor who, when I was nineteen years old, asked me a question that has remained with me ever since: How can one understand another person's pain?

Notes

CHAPTER 1: *The Disappearance of Brian Welsh*

2 *"ice cream with preservatives"*: Testimonials from MCS community, Facebook.

2 *"the list could grow much longer"*: Environmental Sensitivities Resource Team, "Chapter Two: What Causes MCS?," www.mcsresearch.net/chapter -2-what-causes-mcs (accessed November 23, 2019).

3 *"but could include almost anything"*: Pamela Reed Gibson, "Living on the Margins with Access Denied," *Ecopsychology* 9, no. 2 (June 2017): 53–54.

3 *"Hanging mail from a clothesline"*: Michael Fumento, "Sick of It All," Reason.com, June 1996, reason.com/archives/1996/06/01/sick-of-it-all (accessed November 23, 2019).

3 *"Shaving facial hair"*: No author listed, "The VOCless Guide for People with Multiple Chemical Sensitivities and MCS Related Illness," 38, www .riveroflifemedical.org/uploads/2/4/2/8/24285565/vocless_guide.pdf (accessed November 23, 2019).

3 *"Yanking out teeth"*: Linda Sepp, "Invisible Barriers, Invisible Disabilities, Invisible People," *Ecopsychology* 9, no. 2 (June 2017): 68.

3 *"convinced her gynecologist to conduct his examination in the backseat"*: Pamela Reed Gibson, "Living on the Margins with Access Denied," *Ecopsychology*: 54.

3 *"nine years without clean sheets"*: Linda Sepp, "Invisible Barriers, Invisible Disabilities, Invisible People," *Ecopsychology*, 70.

3 *"The phenomenon went by many names"*: YourCare Medical Policy #2.01.04 regarding Clinical Ecology/Multiple Chemical Sensitivities/Idiopathic Environmental Intolerance, yourcarehealthplan.com/Portals/0/PDFs/CMPS /Allergen Immunotherapy 2.01.11.pdf (accessed November 22, 2019).

3 *"I had found Environmental Illness . . . to be the most impartial and inclusive"*: Stanley M. Caress, PhD and Anne C. Steinemann, PhD, "Prevalence of Multiple Chemical Sensitivities: A Population-Based Study in the Southeastern United States," *American Journal of Public Health* 94, no. 5 (June 2004): 746–47.

3 *"First recognized in 1962"*: William J. Rea, "History of chemical sensitivity and diagnosis," *Review of Environmental Health* 31, no. 3 (July 2015): 353.

3 *"believed to afflict as many as 42 million people nationwide"*: Anne Steinemann, "National Prevalence and Effects of Multiple Chemical Sensitivities," *Journal of Occupational and Environmental Medicine* 60, no. 3 (March 2018): 152.

3 *"prevalence had increased over 300 percent"*: Anne Steinemann, "National Prevalence and Effects of Multiple Chemical Sensitivities," *Journal of Occupational and Environmental Medicine,* 155.

3 *"as much as 30 percent of the population experiencing some level of hypersensitivity"*: S. J. Genuis, "Sensitivity-Related Illness: The Escalating Pandemic of Allergy, Food Intolerance and Chemical Sensitivity," *Science of the Total Environment* 408, no. 24, (November 2010): 6048.

3 *"it commenced with a single massive toxic exposure"*: "Chemical Sensitivity: Pathophysiology or Pathopsychology?," S. J. Genuis, *Clinical Therapeutics* 35, no. 5 (May 2013): 573.

3 *"treatments ranging from nutritional supplements to exorcism"*: Alison Johnson, "Alison Johnson's White Paper Concerning Snowflake, Arizona, and the Media," Privately distributed, July 2016.

3 *"houses wallpapered in tinfoil"*: Stephan Bodian, interview by author, August 3, 2016.

3 *"thirty-year-old Benzes with gutted electrical systems"*: Tay Wiles, "In the Middle of Nowhere, a Promised Land," *High Country News*, July 23, 2015, www.hcn.org/articles/environmental-illness-in-arizona (accessed November 23, 2019).

3 *"would only venture out wearing an industrial-grade respirator"*: Judd Robertson, interview by author, August 1, 2016.

3 *"bought and sold four houses"*: Stephan Bodian, interview by author, August 3, 2016.

3   *"EI had cost him $850,000"*: Earle McQuaide, interview by author, August 11, 2016.

4   *"fleeing from one oasis to the next"*: The EI Wellspring, "Life in a Less-Toxic and Solar Powered Travel Trailer," www.eiwellspring.org/saferh/TrailerLife .htm (accessed November 23, 2019).

4   *"'the new refugees'"*: MaryFrances Platt, "The New Refugees," *Ragged Edge*, March/April 2003, www.raggededgemagazine.com/0303/0303ft4.html (accessed November 23, 2019).

4   *"'runners'"*: Delaney Hall, "Snowflake: Episode 123," 99% Invisible, July 15, 2014, 99percentinvisible.org/episode/snowflake/ (accessed November 23, 2019).

4   *"240 places in eight years"*: Daniel Berman, interview by author, August 1, 2016.

4   *"halfway up Mount McKinley"*: Steve McManamon, interview by author, August 1, 2016.

4   *"'It's almost like being a pregnant woman'"*: David Reeves, interview by author, August 1, 2016.

4   *"everything from immunological dysregulation to schizophrenia"*: Stephen Barrett, M.D., "A Close Look at 'Multiple Chemical Sensitivity,'" 1998, 6, 19, 24, 28, www.quackwatch.org/01QuackeryRelatedTopics/mcs.pdf (accessed November 23, 2019).

4   *"often associated with anxiety and depression"*: Michael B. Lax, M.D., and Paul K. Henneberger, "Patients with Multiple Chemical Sensitivities in an Occupational Health Clinic: Presentation and Follow-Up," *Archives of Environmental Health* 50, no. 6 (November/December 1995): 429.

4   *"whether it was a result . . . or a cause was unclear"*: D. Johnson and I. Colman, "The Association Between Multiple Chemical Sensitivity and Mental Illness: Evidence from a Nationally Representative Sample of Canadians," *Journal of Psychosomatic Research* 99, nos. 40–44 (August 2017): 40.

4   *"greater than could be found in other chronic diseases"*: R. M. Bloch and W. J. Meggs, "Comorbidity Patterns of Self-Reported Chemical Sensitivity, Allergy, and Other Medical Illnesses with Anxiety and Depression," *Journal of Nutritional & Environmental Medicine*, 16, 136–48.

4   *"limited to Western industrialized countries"*: S. Bornschein, "Idiopathic Environmental Intolerances (Formerly Multiple Chemical Sensitivity) Psychiatric Perspectives," *Journal of Internal Medicine* 250, no. 4 (October 2001): 312.

4   *"headquarters was renovated and new carpeting installed"*: Claudia S. Miller, "Mechanisms of Action of Addictive Stimuli: Toxicant-Induced Loss of Tolerance," *Addiction* 96, no. 1 (July 2000): 123.

5   *"'And that's your life'"*: David Reeves, interview by author, August 1, 2016.

5   *"more prevalent among adults, less among youth and seniors"*: Dylan Johnson and Ian Colman, "The Association Between Multiple Chemical Sensitivity and Mental Illness: Evidence from a Nationally Representative Sample of Canadians," *Journal of Psychosomatic Research* 99, nos. 40–44 (August 2017): 41.

5   *"more prevalent among women"*: Dylan Johnson and Ian Colman, "The Association Between Multiple Chemical Sensitivity and Mental Illness: Evidence from a Nationally Representative Sample of Canadians," *Journal of Psychosomatic Research,* 41.

5   *"the same could be said of . . . myalgic encephalomyelitis"*: Peggy Rosati Allen, "Chronic Fatigue Syndrome: Implications for Women and Their Health Care Providers During the Childbearing Years," *Journal of Midwifery and Women's Health* 53, no. 4, (July/August 2008): 290.

5   *"the same could be said of . . . fibromyalgia"*: National Fibromyalgia Association, "Prevalence," www.fmaware.org/about-fibromyalgia/prevalence/ (accessed November 23, 2019).

5   *"the same could be said of . . . rheumatoid arthritis"*: T. K. Kvien, "Epidemiological Aspects of Rheumatoid Arthritis: The Sex Ratio," *Annals of the New York Academy of Sciences* 1069 (June 2006): 213.

5   *"the same could be said of . . . chronic Lyme disease"*: Gary P. Wormser, M.D., and Eugene D. Shapiro, M.D., "Implications of Gender in Chronic Lyme Disease," *Journal of Women's Health* 18, no. 6, (June 2009): 832.

5   *"neither the American Medical Association nor the Centers for Disease Control and Prevention"*: Arizona Center for Advanced Medicine, "Multiple Chemical Sensitivity (MCS)," June 26, 2013, www.arizonaadvancedmedi cine.com/Articles/2013/June/Multiple-Chemical-Sensitivity-MCS-.aspx (accessed November 23, 2019).

5   *"it was at best quasi-real"*: Michael K. Magill, M.D. and A. Saruda, "Multiple Chemical Sensitivity Syndrome," *American Family Physician* 58, no. 3 (September 1998): 725; AAAAI Board of Directors, "Idiopathic Environmental Intolerances," *Journal of Allergy and Clinical Immunology* 103 (January 1999): 38; ISRTP Board, "Report of the ISRTP Board," *Regulatory Toxicology and Pharmacology* 18, no. 1 (August 1993): 79; www.osha.gov/SLTC /multiplechemicalsensitivities/index.html (accessed November 23, 2019).

5   *"'the divorce disease'"*: Erik Johnson, Erik on Avoidance: Writings About Mold Avoidance 2000–2015 (Ann Arbor: Paradigm Change, 2015), 443, par adigmchange.me/wp-content/uploads/2019/10/Erik-on-Avoidance-Book -PDF-1.pdf (accessed November 23, 2019).

5 *"he often wished he'd gotten cancer instead"*: Daniel Berman, interview by author, August 11, 2016.

5 *"EI could not be dismissed as merely 'psychogenic' "*: No author listed, "Multiple Chemical Sensitivity: A 1999 Consensus," *Archives of Environmental Health* 54, no. 3 (May/June 1999): 147.

5 *"forbidding air fresheners, perfumes, scented hand lotion"*: Centers for Disease Control and Prevention, "Indoor Environmental Quality Policy," June 2009, nebula.wsimg.com/b91bf0e7fae3e446d37f352ff24f82d0?AccessKey Id=5D08F679D61730E5CF3A&disposition=0&alloworigin=1 (accessed November 23, 2019).

7 *"a pederast, a meth dealer, a thief"*: Confidential source, interview by author, January 19, 2017.

7 *"trailing a cloud of perfume"*: Confidential source, interview by author, August 5, 2016.

7 *"newsprint made her ass itch"*: Confidential source, interview by author, July 28, 2016.

7 *"caused by a passing blimp"*: Confidential source, interview by author, July 25, 2016.

CHAPTER 2: *On the Run*

14 *"crayons laced with asbestos"*: Levin, Myron, and Silverstein Stuart, "Asbestos Found in Imported Crayons and Toy Fingerprint Kits," FairWarning: News of Public Health, Consumer and Environmental Issues, July 8, 2015, www.fair warning.org/2015/07/asbestos-in-toys/ (accessed November 23, 2019).

14 *"tampons doused with weed killer"*: ANO "TV-news," "Organic Tampons and Sanitary Pads Recalled After Traces of Weed Killer Discovered," RT News, February 26, 2016, www.rt.com/news/333665-toxic-tampons-shop -recall/ (accessed November 23, 2019).

14 *"whiskey tainted by . . . antifreeze"*: Alexander C. Kaufman, "Fireball Whisky Recalled in 3 Countries over Antifreeze Ingredient," Huffington Post, October 29, 2014.

14 *"Phthalates in your printer ink"*: Hirohisa Takano, "Environmental Pollution and Allergies," *Journal of Toxologic Pathology* 30, no. 3 (July 2017): 194.

14 *"Formaldehyde in your furniture"*: Alissa Cordner, *Toxic Safety: Flame Retardants, Chemical Controversies, and Environmental Health* (New York: Columbia University Press, 2016), 6.

14 *"Glyphosate in your Cheerios"*: "Glyphosate: Unsafe on Any Plate: Food

Testing Results and Scientific Reasons for Concern," Food Democracy Now and the Detox Project, 1, usrtk.org/wp-content/uploads/2016/11/FDN_Glyphosate_FoodTesting_Report_p2016-3.pdf (accessed November 23, 2019).

14 *"Chlorine in your sex toys"*: Kate Sloan, "It's Surprisingly Hard to Ban Toxic Sex Toys, but Here's How to Protect Yourself," *Glamour*, October 13, 2017, www.glamour.com/story/protecting-yourself-from-toxic-sex-toys (accessed November 23, 2019).

14 *"Triclosan in your underwear"*: Rick Smith and Bruce Lourie, *Slow Death by Rubber Duck: The Secret Danger of Everyday Things* (Berkeley: Counterpoint, 2011), 258.

14 *"toxic scented candles"*: No author listed, "Burning Candles Could Actually Be Toxic for You," *Elle* Australia, August 21, 2017, www.elle.com.au/beauty/burning-candles-could-actually-be-toxic-for-you-14126 (accessed November 23, 2019).

15 *"Subway sandwiches was 'linked' to asthma'*: Deena Shanker, "The Yoga-Mat Chemical's Quiet Fast-Food Exit," *Bloomberg*, August 8, 2016.

15 *"sperm counts among men had fallen"*: Nicola Davis, "Sperm Counts Among Western Men Have Halved in Last 40 Years—Study," *The Guardian*, July 25, 2017, www.theguardian.com/lifeandstyle/2017/jul/25/sperm-counts-among-western-men-have-halved-in-last-40-years-study (accessed November 23, 2019).

15 *"'chemicals in commerce'"*: Maya Salam, "Sperm Count in Western Men Has Dropped over 50 Percent Since 1973, Paper Finds," *The New York Times*, August 16, 2017, www.nytimes.com/2017/08/16/health/male-sperm-count-problem.html (accessed November 23, 2019).

15 *"chemicals in America's favorite kids' food"*: Roni Caryn Rabin, "The Chemicals in Your Mac and Cheese," *The New York Times*, July 12, 2017, www.nytimes.com/2017/07/12/well/eat/the-chemicals-in-your-mac-and-cheese.html (accessed November 23, 2019).

15 *"toxic chemicals in the drinking water"*: No author listed, "Toxic Chemicals in Drinking Water for 6M Americans," Fox News, August 10, 2016, www.foxnews.com/health/toxic-chemicals-in-drinking-water-for-6m-americans (accessed November 23, 2019).

15 *"linked to hormone-mimicking phthalates"*: Smith and Lourie, *Slow Death by Rubber Duck: The Secret Danger of Everyday Things*, 35.

15 *"I could expect one more year of life"*: Andrew Noymer, "Life Expectancy in the USA, 1900–98," u.demog.berkeley.edu/~andrew/1918/figure2.html (accessed November 23, 2019).

15 *"I could expect thirty-two"*: Geoba.se, "The World: Life Expectancy (2019)," www.geoba.se/population.php?pc=world&type=17 (accessed November 23, 2019).

16 *"spending somewhere between 95 and 99 percent of our time indoors"*: Craig Chalquist, "A Look at the Ecotherapy Research Evidence," *Ecopsychology* 1, no. 2 (August 2009): 1.

16 *"85,000 synthetic chemicals in play"*: Jerrold J. Heindel, PhD., et al., "Metabolism Disrupting Chemicals and Metabolic Disorders," *Reproductive Toxicology* 68 (March 2017): 11.

16 *"seventeen pesticides we were . . . exposed to"*: Lourie and Smith, *Toxin Toxout: Getting Harmful Chemicals Out of Our Bodies and Our World*, 58.

16 *"nine thousand food additives"*: Kimberly Kindy, "Food Additives on the Rise as FDA Scrutiny Wanes," *Washington Post*, August 17, 2014, www.washingtonpost.com/national/food-additives-on-the-rise-as-fda-scrutiny-wanes/2014/08/17/828e9bf8-1cb2-11e4-ab7b-696c295ddfd1_story.html (accessed November 23, 2019).

16 *"Canadian children were born pre-polluted"*: Lourie and Smith, *Toxin Toxout: Getting Harmful Chemicals Out of Our Bodies and Our World*, 3.

16 *"3.9 billion pounds of toxic chemicals were dumped"*: Environmental Protection Agency, "2015 TRI National Analysis: Executive Summary," www.epa.gov/sites/production/files/2017-01/documents/tri_na_2015_executive_summary.pdf (accessed November 23, 2019).

16 *"chemical industry itself grew from $1.8 trillion"*: Statista, "Total Revenue of the Global Chemical Industry from 2002 to 2018 (in Billion U.S. Dollars)," www.statista.com/statistics/302081/revenue-of-global-chemical-industry/ (accessed November 23, 2019).

16 *"8.3 billion metric tons of it"*: Katelyn Newman, "Study: Enough Plastic on Earth to Bury Manhattan," *US News and World Report*, July 20, 2017, www.usnews.com/news/world/articles/2017-07-20/theres-enough-unrecycled-plastic-on-earth-to-bury-manhattan-two-miles-deep (accessed November 23, 2019).

16 *"Thyroid and liver cancer up 300 percent"*: Amanda Onion, "Thyroid Cancer Rates Triple, and Scientists Look for Cause," *Live Science*, March 31, 2017, www.livescience.com/58489-thyroid-cancer-rates-tripled.html (accessed November 23, 2019).

16 *"non-Hodgkin's lymphoma"*: American Cancer Society, "Non-Hodgkin Lymphoma Risk Factors," www.cancer.org/cancer/non-hodgkin-lymphoma/causes-risks-prevention/risk-factors.html (accessed November 23, 2019).

16 *"kidney cancer up 200 percent"*: Wong-Ho Chow, "Epidemiology and Risk Factors for Kidney Cancer," *National Review of Urology*, May 2010.

16 *"significant increases in acute myeloid leukemia"*: Phyllis A. Wingo et al., "Long-Term Trends in Cancer Mortality in the United States, 1930–1998," *Cancer* 97, no. 12 (June 2003): 3140.

16 *"significant increases in . . . testicular cancer"*: Katherine A. McGlynn, "Etiologic Factors in Testicular Germ Cell Tumors," *Future Oncology* 5, no. 9 (November 2009): 2.

16 *"significant increases in . . . breast cancer"*: Whiteman, Honor, "Scientists Identify 'High-Priority' Chemicals That May Cause Breast Cancer," *Medical News Today*, May 12, 2014, www.medicalnewstoday.com/articles/276702.php (accessed November 23, 2019); National Institute of Environmental Health Services, "Breast Cancer Risk and Environmental Factors," November 2018, www.niehs.nih.gov/health/materials/environmental_factors_and_breast _cancer_risk_508.pdf (accessed November 23, 2019).

17 *"Autism, for instance"*: Jessica Wright, "The Real Reasons Autism Rates Are up in the U.S.," *Scientific American*, March 3, 2017.

17 *"toxins suspected as likely contributors"*: Cynthia D. Nevison, "A Comparison of Temporal Trends in United States Autism Prevalence to Trends in Suspected Environmental Factors," *Environmental Health* 13, no. 73 (September 2014): 1.

17 *"mental retardation or intellectual impairment (up 63 percent . . .)"*: Amy J. Houtrow et al., "Changing Trends of Childhood Disability, 2001–2011," *Pediatrics* 134, no. 3 (September 2014): 534.

17 *"allergies (sensitization rates approaching 50 percent . . .)"*: American Academy of Allergy, Asthma and Immunology, "Allergy Statistics," www.aaaai.org /about-aaaai/newsroom/allergy-statistics (accessed November 23, 2019).

17 *"obesity (highest prevalence in recorded history . . .)"*: Jerrold J. Heindel, PhD., et al., "Metabolism Disrupting Chemicals and Metabolic Disorders," *Reproductive Toxicology,* 4.

17 *"genital birth defects (increasing)"*: Alisa L. Rich et al., "The Increasing Prevalence in Intersex Variation from Toxicological Dysregulation in Fetal Reproductive Tissue Differentiation and Development by Endocrine-Disrupting Chemicals," *Environmental Health Insights* 10 (September 2016): 163.

17 *"an 'epidemic of male reproductive problems'"*: Russ Hauser et al., "Male Reproductive Disorders, Diseases, and Costs of Exposure to Endocrine-Disrupting Chemicals in the European Union," *Journal of Clinical Endocrinological Metabolism* 100, no. 4 (April 2015): 1268.

17 *"prevalence actually seems to be decreasing"*: World Health Organization

Mortality Database, Search Parameters: Indicators: "No. of Deaths— Alzheimer and Other Dementias, Both Sexes"; Countries: "United States of America"; Years: "1979–2016," apps.who.int/healthinfo/statistics/mor tality/whodpms/ (accessed November 23, 2019); Kenneth M Langa, "Is the Risk of Alzheimer's Disease and Dementia Declining?,"*Alzheimer's Research & Therapy* 7, no. 34 (March 2015): 1.

17   *"scientists . . . suspect that chemicals may play a role"*: Annie Sneed, "DDT, Other Environmental Toxins Linked to Late-Onset Alzheimer's Disease." *Scientific American*, February 10, 2014.

17   *"asthma, the prevalence of which, while increasing, is not on the same order as autism"*: Cynthia D. Nevison, "A Comparison of Temporal Trends in United States Autism Prevalence to Trends in Suspected Environmental Factors," *Environmental Health*, 13.

17   *"'There is just very limited data'"* Amanda Onion, "Thyroid Cancer Rates Triple, and Scientists Look for Cause." Live Science, March 31, 2017, www .livescience.com/58489-thyroid-cancer-rates-tripled.html (accessed November 23, 2019).

18   *"'a vast human experiment'"*: McKay Jenkins, *What's Gotten into Us?: Staying Healthy in a Toxic World* (New York: Random House, 2011), 12; Zoë Schlanger, "A New Documentary Probes the Vast Human Experiment of Unregulated Chemicals," *Newsweek*, April 17, 2015; Jaakko Lyytinen, "The World's Largest-Ever Human Experiment." The Electrosensitive Society, March 4, 2010, www.electrosensitivesociety.com/2010/04/08/the-worlds -largest-ever-human-experiment/ (accessed November 23, 2019); John Molot, M.D., "A Human Experiment?," *Environmental Medicine*, May 4, 2016, johnmolot.com/2016/05/04/a-human-experiment/ (accessed November 23, 2019); No author listed, "Doctor Jekyll and Mr Formaldehyde," Cage Free Zen: A Virtual Whiff of Multiple Chemical Sensitivity, April 17, 2011, cagefreezen.wordpress.com/2011/04/17/doctor-jekyll-and-mr-formal dehyde/ (accessed November 23, 2019).

18   *"permits the introduction of hundreds of untested new chemicals"*: Kimberly Kindy, "Food Additives on the Rise as FDA Scrutiny Wanes," *Washington Post*, August 17, 2014, www.washingtonpost.com/national/food-additives-on -the-rise-as-fda-scrutiny-wanes/2014/08/17/828e9bf8-1cb2-11e4-ab7b-69 6c295ddfd1_story.html (accessed November 23, 2019); "Chemicals, Cancer, and You," Agency for Toxic Substances and Disease Registry, 2009, 1, www.atsdr.cdc.gov/emes/public/docs/Chemicals,%20Cancer,%20and%20 You%20FS.pdf (accessed November 23, 2019).

18   *"assumed to be safe until proven otherwise"*: Nicholas Kristof, "Contaminating Our Bodies with Everyday Products," *The New York Times*, November 29, 2015, www.nytimes.com/2015/11/29/opinion/sunday/contaminating-our-bodies-with-everyday-products.html (accessed November 23, 2019).

18–19 *"85,000+ chemicals in our environment"*: Jerrold J. Heindel, PhD., et al., "Metabolism Disrupting Chemicals and Metabolic Disorders," *Reproductive Toxicology*, 11.

19   *"never been tested either individually or in combination"*: McKay Jenkins, *What's Gotten into Us?: Staying Healthy in a Toxic World* (New York: Random House, 2011), 13, 213; Alissa Cordner, *Toxic Safety: Flame Retardants, Chemical Controversies, and Environmental Health* (New York: Columbia University Press, 2016), 6.

19   *"Not even children's toys were required to be independently tested"*: www.cpsc.gov/Business--Manufacturing/Testing-Certification/Lab-Accreditation/CPSC-Accredited-Firewalled-Laboratories (accessed November 23, 2019).

19   *"'Responsible companies test their products'"*: United States Consumer Product Safety Division, "CPSC-Accredited Firewalled Laboratories," www.cpsc.gov/About-CPSC/Contact-Information (accessed November 23, 2019).

19   *"($100,000 per member . . .)"*: OpenSecrets.org: Center for Responsive Politics, "Top Contributors, 2019-2020," www.opensecrets.org/industries/indus.php?ind=N13 (accessed November 23, 2019).

19   *"'put the regulations industry out of work'"*: Eric Lipton, "Trump Campaigned Against Lobbyists, but Now They're on His Transition Team," *The New York Times*, November 12, 2016, www.nytimes.com/2016/11/12/us/politics/trump-campaigned-against-lobbyists-now-theyre-on-his-transition-team.html (accessed November 23, 2019).

19   *"removing a ban on lobbyists"*: Eric Lipton, interview by Dave Davies, *Fresh Air*, NPR, August 16, 2017, www.npr.org/2017/08/16/543876454/in-trumps-government-the-regulated-have-become-the-regulators (accessed November 23, 2019).

19   *"appointing a known climate change denier"*: Coral Davenport and Eric Lipton, "Trump Picks Scott Pruitt, Climate Change Denialist, to Lead E.P.A.," *The New York Times*, December 7, 2016, www.nytimes.com/2016/12/07/us/politics/scott-pruitt-epa-trump.html (accessed November 23, 2019).

19   *"the director of eight different pharmaceutical companies"*: Eric Lipton, Ben Protess, and Andrew W. Lehren, "With Trump Appointees, a Raft of

Potential Conflicts and 'No Transparency,'" *The New York Times,* April 15, 2017, www.nytimes.com/2017/04/15/us/politics/trump-appointees-potential -conflicts.html (accessed November 23, 2019).

19 *"could actually make you less healthy"*: Deena Shanker, "How Organic Pro- duce Can Make America Less Healthy," *Bloomberg,* March 9, 2017, www .bloomberg.com/news/articles/2017-03-09/how-organic-produce-can-make -america-less-healthy (accessed November 23, 2019).

19 *"(lined with BPA plastic)"*: Tony Iallonardo, "Report Finds Toxic BPA Com- mon in Food Cans," Safer Chemicals, Healthy Families, March 30, 2016, saferchemicals.org/2016/03/30/12949/ (accessed November 23, 2019).

20 *"(the product of 275 different volatile organic compounds)"*: Smith and Lourie, *Slow Death by Rubber Duck: The Secret Danger of Everyday Things,* 156.

20 *"25 percent increase in books on anxiety"*: Lauren Thomas, "Barnes & Noble Says Sales of Books Related to Anxiety Are Soaring. Here's Why," CNBC, August 1, 2018, www.cnbc.com/2018/08/01/barnes--noble-says-sales-of -books-related-to-anxiety-are-soaring.html (accessed November 23, 2019).

20 *"bottled air from Banff"*: Alex Moshakis, "Fresh Air for Sale," *The Guard- ian,* January 21, 2018, www.theguardian.com/global/2018/jan/21/fresh-air- for-sale (accessed November 23, 2019).

20 *"impacted four times as many people"*: Centers for Disease Control and Prevention, "Myalgic Encephalomyelitis/Chronic Fatigue Syndrome," www .cdc.gov/me-cfs/index.html;   www.cdc.gov/arthritis/basics/fibromyalgia.htm (accessed November 23, 2019).

20 *"'Twentieth-Century Disease'"*: Glenna Whitley, "Is the 20th Century Mak- ing You Sick?," D Magazine, August 1990, www.dmagazine.com/publications /d-magazine/1990/august/is-the-20th-century-making-you-sick/ (accessed November 23, 2019); Eric Liskey, "Part II: Multiple Chemical Sensitivity," Grounds Maintenance, grounds-mag.com/mag/grounds_maintenance_part _ii_multiple/ (accessed November 23, 2019); P. Bartsch, "Environmental Diseases, Diseases of the 21st Century? Multiple Chemical Hypersensitiv- ity," *Revue Medicale de Liege* 51, no. 8, (August 1996): 527.

CHAPTER 3: *The Comfort of Guns*

22 *"clearly stipulated cause and victim"*: Geneviève Nadeau and Katherine Lippel, "From Individual Coping Strategies to Illness Codification: The Reflection of Gender in Social Science Research on Multiple Chemical

Sensitivities (MCS)," *International Journal for Equity in Health* 13, no. 78 (September 2014): 3.

22 *"there was no such thing"*: S. J. Genuis, "Sensitivity-Related Illness: The Escalating Pandemic of Allergy, Food Intolerance and Chemical Sensitivity," *Science of the Total Environment,* 6053.

22 *"dozens of biomarkers had been identified"*: D. Belpomme, C. Campagnac, and P. Irigaray, "Reliable Disease Biomarkers Characterizing and Identifying Electrohypersensitivity and Multiple Chemical Sensitivity as Two Etiopathogenic Aspects of a Unique Pathological Disorder," *Reviews on Environmental Health*, 30 (2015), 251–71; L. Andersson et al., "Chemosensory Perception, Symptoms and Autonomic Responses During Chemical Exposure in Multiple Chemical Sensitivity," *International Archives of Occupational and Environmental Health* 89, no. 1 (January 2016): 80; S. J. Genuis, "Sensitivity-Related Illness: The Escalating Pandemic of Allergy, Food Intolerance and Chemical Sensitivity," *Science of the Total Environment,* 6053.

23 *"none were found in* all *sensitives"*: L. Andersson et al., "Chemosensory Perception, Symptoms and Autonomic Responses During Chemical Exposure in Multiple Chemical Sensitivity", *International Archives of Occupational and Environmental Health*, 80; S. J. Genuis, "Sensitivity-Related Illness: The Escalating Pandemic of Allergy, Food Intolerance and Chemical Sensitivity," *Science of the Total Environment,* 6053;

23 *"if they all derived from a common underlying source"*: Pamela Reed Gibson, "Living on the Margins with Access Denied," *Ecopsychology*, 56.

23 *"sensitives' sense of smell"*: "Chemosensory Perception, Symptoms and Autonomic Responses During Chemical Exposure in Multiple Chemical Sensitivity," *International Archives of Occupational and Environmental Health*: 80.

23 *"funding for research on the health of Gulf War veterans totaled more than $500 million"*: Veterans Administration. "Gulf War Research Strategic Plan 2013–2017, 2015 Update," Washington, DC: Veterans Health Administration; 2015b, 14, www.research.va.gov/pubs/docs/GWResearch-StrategicPlan.pdf (accessed November 23, 2019).

23 *"shown to exhibit structural differences"*: Kimberly Sullivan, "Structural MRI and Cognitive Correlates in Pest-Control Personnel from Gulf War I," Defense Technical Information Center Report, Prepared for U.S. Army Medical Research and Materiel, Fort Detrick, Maryland: 30 apps.dtic.mil/docs/citations/ADA501573 (accessed November 23, 2019).

24 *"encouraging researchers to move on"*: Deborah Cory-Slechta and Roberta

Wedge, *Gulf War and Health: Volume 10: Update of Health Effects of Serving in the Gulf War, 2016* (Washington, D.C.: National Academies Press, 2016), 5–6.

24 *"headaches, too, lacked reliable biomarkers"*: Paul Rizzoli, MD, William J. Mullally, MD, "Headache," *The American Journal of Medicine* 131, no. 1 (January 2018): 19.

24 *"role of the hypothalamus in cluster headaches"*: Alina Buture et al., "Current Understanding on Pain Mechanism in Migraine and Cluster Headache," *Anesthesiology and Pain Medicine* 6, no. 3 (June 2016): 4.

24 *"the role of the brain stem in migraines"*: Alina Buture et al., "Current Understanding on Pain Mechanism in Migraine and Cluster Headache," *Anesthesiology and Pain Medicine* 6, no. 3 (June 2016): 3.

24 *"MTHFR gene in migraines with aura"*: Innocenzo Rainero et al., "Genes and Primary Headaches: Discovering New Potential Therapeutic Targets," *The Journal of Headache and Pain* 14, no. 61 (2013): 2.

24 *"the psychogenic explanation had taken a backseat"*: Nauman Tariq, interview by author, November 2, 2017.

24 *"huge explosion in pain studies"*: www.ncbi.nlm.nih.gov/pubmed search: "pain": from 6,281 studies in 1985 to 43,820 studies in 2018.

24 *"ballooning interest in chronic pain"*: Dan Clauw, interview by author, November 1, 2017; Nicola Twilley, "The Neuroscience of Pain," *The New Yorker*, July 2, 2018.

24 *"affected over 25 million Americans"*: R. L. Nahin, "Estimates of Pain Prevalence and Severity in Adults: United States, 2012," *Journal of Pain* 16, no. 8 (2015): 769.

24 *"allowed chronic pain patients to be recognized as victims"*: Haider Warraich, "Is Pain a Sensation or an Emotion?," *The New York Times*, March 16, 2019, www.nytimes.com/2019/03/16/opinion/sunday/pain-opioids.html (accessed November 23, 2019).

24 *"now generously referred to as 'centralized'"*: Dan Clauw, interview by author, November 1, 2017.

24 *"Somatoform disorder . . . had itself been redefined"*: Morton E. Tavel, M.D., "Somatic Symptom Disorders Without Known Physical Causes- One Disease with Many Names?," *The American Journal of Medicine* 128, no. 10 (October 2015): 1054.

24 *"women experienced pain differently than men"*: E. J. Bartley and R. B. Fillingim, "Sex Differences in Pain: A Brief Review of Clinical and Experimental Findings," *British Journal of Anaesthesia* 52, no. 8 (July 2013): 52.

25 *"patients were now being accorded greater credibility"*: Dan Clauw, interview by author, November 1, 2017.

25 *"whose playbook the chemical industry zealously followed"*: McKay Jenkins, *What's Gotten into Us?: Staying Healthy in a Toxic World*, 129; Jennifer Sass, "Toxic Chemical Industry and House R's Attack on Science," NRDC, September 7, 2017 www.nrdc.org/experts/jennifer-sass/toxic-chemical -industry-and-house-rs-attack-science (accessed November 23, 2019); Nicholas Kristof, "Trump's Legacy: Damaged Brains." *The New York Times*, October 28, 2017, www.nytimes.com/interactive/2017/10/28/opinion/sunday /chlorpyrifos-dow-environmental-protection-agency.html (accessed November 23, 2019).

CHAPTER 4: *Crazy Charlie*

33 *"'Once invited, the terror begins'"*: Multiple Chemical Survivor, "Open Letter to the Snowflakers, or How to Avoid Looking Crazy," Multiple Chemical Survivor: Living and Surviving with Multiple Chemical Sensitivity, July 26, 2016, multiplechemicalsurvivor.blogspot.com/2016/07/open-letter-to-snow flake-community-or.html (accessed November 23, 2019).

34 *"original explanation for EI emerged from an allergy model"*: Herman Staudenmayer, *Environmental Illness: Myth and Reality* (Boca Raton: CRC Press, 1998), 13.

35 *"a theory that came to be known as neural sensitization"*: I. R. Bell, "White paper: Neuropsychiatric Aspects of Sensitivity to Low-Level Chemicals: A Neural Sensitization Model." *Toxicology and Industrial Health* 10, nos. 4–5, July/October 1994: 281.

35 *"'neurogenic inflammation'"*: W. J. Meggs, "Neurogenic Inflammation and Sensitivity to Environmental Chemicals," *Environmental Health Perspectives* 101, no. 3 (August 1993): 234.

35 *"mustard oil squirted into the eyes of rabbits"*: Peter Baluk, "Neurogenic Inflammation in Skin and Airways," *Journal of Investigative Dermatology* 2, no. 1 (August 1997): 76.

35–36 *"a phenomenon known as 'spreading'"*: Laurie Dennison Busby, "A Comparison of Multiple Chemical Sensitivity with Other Hypersensitivity Illnesses Suggests Evidence and a Path to Answers," *Ecopsychology* 9, no. 2 (June 2017): 93.

36 *"the more agents it perceived as threats"*: Steven C. Rowat, "Integrated Defense System Overlaps as a Disease Model: With Examples for Multiple

Chemical Sensitivity," *Environmental Health Perspectives* 106, no. 1 (February 1998): 87.

36 *"The paper was unprecedented in its specificity"*: Martin L. Pall, "Common Etiology of Posttraumatic Stress Disorder, Fibromyalgia, Chronic Fatigue Syndrome and Multiple Chemical Sensitivity via Elevated Nitricoxide/Peroxynitrite," *Medical Hypotheses* 57, no. 2 (August 2001): 139.

36 *"(. . . over twenty-two distinct mechanisms)"*: Martin L. Pall, "Multiple Chemical Sensitivity: Toxicological Questions and Mechanisms," *General and Applied Toxicology* (2009): 12.

37 *"the 'spreading' phenomenon and why sensitives were so sensitive"*: Martin L. Pall, "Multiple Chemical Sensitivity: Toxicological Questions and Mechanisms," *General and Applied Toxicology* (2009): 11.

37 *"the playbook of a deranged football coach"*: Martin L. Pall, "Multiple Chemical Sensitivity: Toxicological Questions and Mechanisms," *General and Applied Toxicology* (2009): 12.

37 *"rarely cited in the literature"*: Martin Pall, interview by author, November 20, 2017.

37 *"shopping malls misted their lobbies with perfume"*: Air Aroma, "Airports to Airlines: Scents That Ensure Passengers Arrive Relaxed and Refreshed," www.air-aroma.com/scenting/airports-and-airlines (accessed November 23, 2019).

CHAPTER 5: *Ignorance Ain't Easy*

42 *"articles that regularly appeared claiming to debunk the mold threat"*: Farah Khan, "Why Is the Internet So Obsessed with 'Toxic Mold'?," The Daily Beast, October 15, 2016, www.thedailybeast.com/why-is-the-internet-so-obsessed-with-toxic-mold (accessed November 23, 2019).

42 *"'That epidemic is now happening'"*: Confidential source, Facebook.

42 *"the toxic chemicals secreted by mold"*: Andrea T. Borchers, Christopher Chang, and M. Eric Gershwin, "Mold and Human Health: A Reality Check," *Clinical Reviews in Allergy and Immunology* 52, no. 3 (June 2017): 309.

43 *"entire villages wiped out"*: J. David Miller, "Mycotoxins in Small Grains and Maize: Old Problems, New Challenges," *Food Additives & Contaminants: Part A* 25, no. 2 (February 2008): 219.

43 *"several others as 'possibly' carcinogenic"*: Andrea T. Borchers, Christopher Chang, and M. Eric Gershwin, "Mold and Human Health: A Reality Check," *Clinical Reviews in Allergy and Immunology* 52, no. 3 (June 2017): 311.

43 *"The Institute of Medicine dismissed the possibility"*: Amanda Rudert and Jay Portnoy, "Mold Allergy: Is It Real and What Do We Do About It?," *Expert Review of Clinical Immunology* 13, no. 8 (December 2017): 827.

43 *"ten times that number had yet to be discovered"*: Andrea T. Borchers, Christopher Chang, and M. Eric Gershwin, "Mold and Human Health: A Reality Check," *Clinical Reviews in Allergy and Immunology* 52, no. 3 (June 2017): 305.

44 *"their own dating service"*: www.canarysingles.com/ (accessed November 23, 2019).

45 *"dismissing non-sensitives as 'normies'"*: Sidney Stevens, "The Residents of this Arizona Community Are Allergic to Modern Life," Mother Nature Network, December 19, 2016, www.mnn.com/health/allergies/stories /residents-Arizona-community-Snowflake-allergic-modern-life (accessed November 23, 2019); Linda Sepp, "There's WHAT in my Baking Soda?," Seriously "Sensitive" to Pollution, March 27, 2013, lindasepp.wordpress. com/2013/03/27/theres-what-in-my-baking-soda/ (accessed November 23, 2019).

45 *"'Chemical AIDS'"*: Chiara De Luca et al., "Biological Definition of Multiple Chemical Sensitivity from Redox State and Cytokine Profiling and Not from Polymorphisms of Xenobiotic-Metabolizing Enzymes," *Toxicology and Applied Pharmacology* 248, no. 3 (November 2010): 286; United States Department of Labor, "Multiple Chemical Sensitivities," www.osha.gov/SLTC /multiplechemicalsensitivities/index.html (accessed November 23, 2019).

45 *"a flea bomb in a drop ceiling"*: Confidential source, interview by author.

45 *"Brain retraining was a commercial application of 'neuroplasticity'"*: S. T. Wilkinson, "Leveraging Neuroplasticity to Enhance Adaptive Learning: The Potential for Synergistic Somatic-Behavioral Treatment Combinations to Improve Clinical Outcomes in Depression," *Biological Psychiatry* 85, no. 6 (March 2019): 454; G. Lubrini, "Brain Disease, Connectivity, Plasticity and Cognitive Therapy: A Neurological View of Mental Disorders," *Neurologia* 33, no. 3 (April 2018): 187.

46 *"'Seems along the same reasoning to me'"*: Confidential source, Facebook.

46 *"Pall ridiculed competing researchers"*: Martin L. Pall, "Multiple Chemical Sensitivity is a Response to Chemicals Acting as Toxicants via Excessive NMDA Activity," *Journal of Psychosomatic Research* 69, no. 3 (September 2010): 327.

46 *"'the old English joke about a drunk'"*: Martin L. Pall, "Multiple Chemical Sensitivity Is a Response to Chemicals Acting as Toxicants via Excessive NMDA Activity," *Journal of Psychosomatic Research*, 327.

47 *"'they divine unfounded beliefs'"*: Herman Staudenmayer, *Environmental Illness: Myth and Reality*, 17.

47 *"a fear of confronting reality"*: Herman Staudenmayer, *Environmental Illness: Myth and Reality*, 17.

47 *"'the only ailment in existence'"*: R.E. Gots, "Multiple Chemical Sensitivities— Public Policy," *Journal of Toxicology: Clinical Toxicology* 33, no. 2 (1995): 111.

47 *"'flat-earthers of the future'"* *"the very definition of grasping-at-straws-failures," "shills for the chemical industry,"* and *"frauds"*: Amazon product review: Stephan J. Barrett, *Chemical Sensitivity: The Truth About Environmental Illness* www.amazon.ca/gp/customer-reviews/R217VTYEP773U2 /ref=cm_cr_arp_d_rvw_ttl (accessed November 23, 2019); Confidential source, Facebook; Will Moredock, "The Best Science Money Can Buy," The Environmental Illness Resource, May 18, 2016, www.ei-resource .org/articles/multiple-chemical-sensitivity-articles/the-best-science-mon ey-can-buy-the-debunking-of-multiple-chemical-sensitivity-mcs/ (accessed November 23, 2019); Patrick Pontillo, "MCS Denier, EPIC Takedown #2: Doctor Jekyll and Mr. Formaldehyde," Labyrinth Press, May 20, 2015, the-labyrinth.com/2015/05/20/doctor-jekyll-and-mr-formaldehyde/ (accessed November 23, 2019).

48 *"threatening to castrate him"*: David Tuller, DrPH, "Trial by Error: The Troubling Case of the PACE Chronic Fatigue Syndrome Study," Virology Blog, October 21, 2015, www.virology.ws/2015/10/21/trial-by-error-i/ (accessed November 23, 2019).

49 *"'Heartsink patients'"*: Tim C. olde Hartman et al., "Medically Unexplained Symptoms, Somatisation Disorder and Hypochondriasis: Course and Prognosis: A Systematic Review," *Journal of Psychosomatic Research* 66, no. 5 (May 2009): 364.

49 *"the disease lacked a biological etiology"*: I. R. Bell, "An Olfactory-Limbic Model of Multiple Chemical Sensitivity Syndrome: Possible Relationships to Kindling and Affective Spectrum Disorders," *Biological Psychiatry* 32, no. 3 (August 1992): 219.

49 *"detox programs and 'botanical medicines' for children with autism"*: Casey Ross, Max Blau, and Kate Sheridan, "Medicine with a Side of Mysticism: Top Hospitals Promote Unproven Therapies," *StatNews*, March 7, 2017, www.statnews.com/2017/03/07/alternative-medicine-hospitals-promote/ (accessed November 23, 2019).

CHAPTER 6: *Down into the Country*

52 *"the demise of the local sugar industry"*: George de Lama, "Sugar King No More in Hawaii," *Chicago Tribune*, June 19, 1994, www.chicagotribune.com /news/ct-xpm-1994-06-19-9406190289-story.html (accessed November 23, 2019).

52 *"Monsanto introduced saccharin as its first product"*: Kathryn Steen, *The American Synthetic Organic Chemicals Industry* (Chapel Hill: The University of North Carolina Press, 2014), 33.

55 *"the most well-known research on EI was that of Claudia Miller"*: Claudia S. Miller, "Toxicant-Induced Loss of Tolerance: An Emerging Theory of Disease?," *Environmental Health Perspectives* 105, Supplement 2: Chemical Sensitivity (March 1997): 445.

55 *"even minimal exposures could provoke outsized effects"*: Claudia S. Miller, "Toxicant-Induced Loss of Tolerance: An Emerging Theory of Disease?," *Environmental Health Perspectives*: 445.

56 *"chemically sensitive patients respond as though they were ab-dicted"*: Claudia S. Miller, "Toxicant-Induced Loss of Tolerance: An Emerging Theory of Disease?," *Environmental Health Perspectives*: 447.

56 *"a sort of 'cryptotoxicity'"*: Claudia S. Miller, "Chemical Sensitivity: Symptom, Syndrome or Mechanism for Disease?" *Toxicology* 111, nos. 1–3 (July 1996): 81.

56 *"She did cite Bell's neural sensitization"*: Claudia S. Miller, "Mechanisms of Action of Addictive Stimuli: Toxicant-Induced Loss of Tolerance," *Addiction*, 126.

57 *"'the vituperative professional disputes that surround [EI]'"*: Claudia S. Miller, "Toxicant-Induced Loss of Tolerance: An Emerging Theory of Disease?," *Environmental Health Perspectives*: 447.

58 *"both proposals were met with a certain degree of outrage"*: Herman Staudenmayer, *Environmental Illness: Myth and Reality*, 298.

59 *"inveighing against 'carpet and rug manufacturers . . .'"*: Claudia S. Miller, "Mechanisms of Action of Addictive Stimuli: Toxicant-Induced Loss of Tolerance," *Addiction*, 132.

59 *"'a less than enthusiastic reception'"*: Claudia S. Miller, "Chemical Sensitivity: Symptom, Syndrome or Mechanism for Disease?," *Toxicology*, 84.

60 *"Pall also compared his to work to Koch's"*: Martin L. Pall, "Multiple Chemical Sensitivity: Toxicological Questions and Mechanisms," *General and Applied Toxicology*, 24.

CHAPTER 7: *The Word from Snowflake*

63   *"'I self-interrupt'"*: Bruce McCreary, interview with Mae Ryan, "Snowflake, Arizona: Where the Residents Are Allergic to Life," *The Guardian*, July 11, 2016, www.theguardian.com/society/2016/jul/11/snowflake-arizona-environ mental-illness (accessed November 23, 2019).

63   *"became chemically sensitive after a workplace accident involving solvents"*: Fred A. Bernstein, "In One Arizona Community, an Oasis in a Toxic World," *The New York Times*, July 10, 2005, www.nytimes.com/2005/07/10/realestat e/in-one-arizona-community-an-oasis-in-a-toxic-world.html (accessed No-vember 23, 2019); J. Appleby, "Built to Be Bare: In Marin County, a HUD-Backed Haven for the Chemically Sensitive." *The Washington Post*, March 2, 1995.

63   *"eventually settled in Snowflake"*: Tay Wiles, "In the Middle of Nowhere, a Promised Land," *High Country News*, July 23, 2015, www.hcn.org/articles /environmental-illness-in-arizona (accessed November 23, 2019).

63   *"Susan Molloy, who became sensitive at age thirty-one"*: Tay Wiles, "In the Middle of Nowhere, a Promised Land," *High Country News*, July 23, 2015, www.hcn.org/articles/environmental-illness-in-arizona (accessed November 23, 2019).

63   *"a complete EI home building protocol"*: The EI Wellspring, "Desert Moon House: Construction of a Healthy House," www.eiwellspring.org /DMH-Construction.htm (accessed November 23, 2019).

64   *"Suicides happened as often as twice a year"*: Kathleen Hale, "Allergic to Life: The Arizona Residents 'Sensitive to the Whole World,'" *The Guardian*, July 11, 2016, www.theguardian.com/society/2016/jul/11/snowflake-arizona -environmental-illness (accessed November 23, 2019).

64   *"People had been known to starve themselves"*: Kathleen Hale, "Allergic to Life: the Arizona Residents 'Sensitive to the Whole World,'" *The Guardian*, July 11, 2016, www.theguardian.com/society/2016/jul/11/snowflake-arizona -environmental-illness (accessed November 23, 2019).

64   *"'I left the protective life that I was living'"*: Confidential source, interview by author, July 25, 2016.

64   *"'similar to writing an article about marriage'"*: Confidential source, email.

68   *"'a scoop of zeolite'"*: Confidential source, Facebook.

CHAPTER 8: *The Work of the Devil*

70 *"'Glockzilla'"*: Mike Searson, "Glock G40 Gen4 – aka Glockzilla, is a 10mm Masterpiece," Ammoland, September 23, 2015, www.ammoland.com/2015 /09/glock-g40-gen4-a-10mm-masterpiece (accessed November 23, 2019).

72 *"higher rates in mental illness that accompanied EI"*: Dylan Johnson and Ian Colman, "The Association Between Multiple Chemical Sensitivity and Mental Illness: Evidence from a Nationally Representative Sample of Canadians," *Journal of Psychosomatic Research,* 43; E. M. Weiss et al., "Differences in Psychological and Somatic Symptom Cluster Score Profiles Between Subjects with Idiopathic Environmental Intolerance, Major Depression and Schizophrenia," *Psychiatry Research* 249 (March 2017): 191; E. Lago Blanco et. al., "Multiple Chemical Sensitivity: Clinical Evaluation of the Severity and Psychopathological Profile," *Medicina Clinica* 146, no. 3 (February 2016): 111; Michael B. Lax, M.D., and Paul K. Henneberger, "Patients with Multiple Chemical Sensitivities in an Occupational Health Clinic: Presentation and Follow-Up," *Archives of Environmental Health* 50, no. 6 (1995): 425.

72 *"other types of chronic illness"*: Pamela Reed Gibson, "Living on the Margins with Access Denied," *Ecopsychology:* 55.

73 *"more likely to perceive images"*: Jennifer A. Whitson and Adam D. Galinsky, "Lacking Control Increases Illusory Pattern Perception," *Science* 322, no. 5898 (October 2008): 115.

74 *"(. . . the smell of a place you don't want to live)"*: Confidential source, Facebook.

74 *"'if you have encountered it, you know it'"*: Confidential source, Facebook.

74 *"assumed to be the work of the devil"*: Cecilia Tasca et al., "Women and Hysteria in the History of Mental Health," *Clinical Practice & Epidemiology in Mental Health* 8 (2012): 2012.

74 *"compare EI to actual witchcraft"*: W. J. Waddell, "Science of Toxicology and its Relevance to MCS," *Regulatory Toxicology and Pharmacology* 18, no. 1 (August 1993): 116.

75 *"'science may deny, or at least cast aside as of no scientific interest'"*: Michael Polanyi, *Personal Knowledge: Towards a Post-Critical Philosophy* (Chicago: University of Chicago Press, 1974), 308.

76 *"'If I'm a kook, OK, at least I feel better'"*: Steve Kroll-Smith and H. Hugh Floyd, *Bodies in Protest: Environmental Illness and the Struggle over Medical Knowledge* (New York: NYU Press, 1997), 137.

76   *"'Do you think this is part of our phenomenon'"*: Confidential source, Facebook.

76   *"'It's like a Jason Bourne movie'"*: Daniel Berman, interview by author, August 1, 2016.

77   *"the list numbered around sixty"*: Elizabeth, Erin, "Unintended Holistic Doctor Death Series: Over 90 Dead," Health Nut News, March 12, 2016, www.healthnutnews.com/recap-on-my-unintended-series-the-holistic-doctor-deaths/ (accessed November 23, 2019).

77   *"'MK Ultra style plus antidepressants'"*: Confidential source, Facebook.

78   *"it could be attributed to the sensitivity of the female character"*: Siri Hustvedt, "Philosophy Matters in Brain Matters," *Seizure* 22, no. 3 (April 2013): 171; Kayla Webley Adler, "Women Are Dying Because Doctors Treat Us Like Men," *Marie Claire*, April 25, 2017, www.marieclaire.com/health-fitness /a26741/doctors-treat-women-like-men/ (accessed November 23, 2019).

78   *"the dozen other possible things that might explain it"*: Pamela Reed Gibson, "Multiple Chemical Sensitivity, Culture, and Delegitimization: A Feminist Analysis," *Feminism and Psychology* 7, no. 4 (November 1997): 475.

78   *"'wine and chocolates'"*: Geneviève Nadeau and Katherine Lippel, "From Individual Coping Strategies to Illness Codification: The Reflection of Gender in Social Science Research on Multiple Chemical Sensitivities (MCS)," *International Journal for Equity in Health* 13, no. 78 (September 2014): 7.

78   *"even cancer was once attributed to their 'greater feebleness'"*: Robert Proctor, *The Cancer Wars*, (New York: Basic Books, 1996), 22.

78–79   *"women were less likely to be taken seriously"*: Anne Wernera and Kirsti Malterud, "It Is Hard Work Behaving as a Credible Patient: Encounters Between Women with Chronic Pain and Their Doctors," *Social Science and Medicine* 57, no. 8 (October 2003): 1409.

79   *"NIH began requiring its research grantees to study both sexes"*: Kayla Webley Adler, "Women Are Dying Because Doctors Treat Us Like Men," *Marie Claire*, April 25, 2017, www.marieclaire.com/health-fitness/a26741/doctors-treat-women-like-men/ (accessed November 23, 2019); *"National Institutes of Health, Report of the Advisory Committee on Research on Women's Health, Fiscal Years 2015–2016,"* (Bethesda: NIH Publication, 2017): orwh .od.nih.gov/sites/orwh/files/docs/ORWH_Biennial_Report_WEB_508_FY -15-16.pdf (accessed November 23, 2019.)

79   *"'Where can I read more about hell toxin'"*: Confidential source, Facebook.

80   *"their whole lives are consumed by this 'hell toxin'"*: Confidential source, email to author.

80 *"vast amounts of ad money"*: Katherine Ellen Foley, "Big Pharma Spent an Additional $9.8 Billion on Marketing in the Past 20 Years. It Worked," Quartz, January 9, 2019, qz.com/1517909/big-pharma-spent-an-additional-9-8-billion -on-marketing-in-the-past-20-years-it-worked/ (accessed November 23, 2019).

CHAPTER 9: *Paiute Forestry*

83 *"She had trained with Dr. Michael R. Gray"*: Michael R. Gray, interview by author, August 13, 2016.

83 *"had seen over sixty thousand patients"*: Michael R. Gray, Michael R. Gray resume: portal.azoah.com/oedf/documents/2017A-EMS-0007-DHS/AMH -27-SLEMS.027.dr.gray's%20bio.pdf (accessed November 23, 2019).

83 *"had trained with the founding father"*: Leif Strickland, "Sick of the World," D Magazine, January 2004, www.dmagazine.com/publications/d-magazine /2004/january/sick-of-the-world/ (accessed November 23, 2019).

83–84 *"'look back with astonishment'"*: S. J. Genuis et al., "Clinical Detoxifica- tion: Elimination of Persistent Toxicants from the Human Body," *Scientific World Journal* (June 2013): 2.

84 *"'scientific censorship is dangerous'"*: S. J. Genuis et al., "Incorporating En- vironmental Health in Clinical Medicine," *Journal of Environmental and Public Health* (May 2012): 2.

84 *"'ionic footbaths'"* ... *"'earthing'"* ... *"'perspiration-moistened animal skins'"*: Deborah A. Kennedy et al., "Objective Assessment of an Ionic Footbath (IonCleanse): Testing Its Ability to Remove Potentially Toxic Elements from the Body," *Journal of Environmental and Public Health* (May 2012): 1; Gaétan Chevalier et al., "Earthing: Health Implications of Reconnecting the Human Body to the Earth's Surface Electrons," *Journal of Environmental and Public Health* (May 2012): 2.

87–88 *"the hottest start to summer ever"*: Santos, Fernanda, "Raging Wildfires in the Southwest Stretch Resources," *The New York Times*, June 22, 2016, www.nytimes.com/2016/06/23/us/raging-wildfires-in-the-southwest-stretch -resources.html (accessed November 23, 2019).

87–88 *"temperatures reaching 120 degrees"*: Loki, Reynard, "The Southwest Is Burning: Wildfires Rage Amidst Record-Breaking Heat Wave," Salon.com, June 22, 2016, www.salon.com/2016/06/22/the_southwest_is_burning_partner/ (accessed November 23, 2019).

88 *"five people had died while hiking"*: Fernanda Santos, "Raging Wildfires in the Southwest Stretch Resources," *The New York Times*, June 22, 2016,

www.nytimes.com/2016/06/23/us/raging-wildfires-in-the-southwest-stretch
-resources.html (accessed November 23, 2019).

88 *"planes were being rerouted"*: Reynard Loki, "The Southwest Is Burning:
Wildfires Rage Amidst Record-Breaking Heat Wave," Salon.com, June 22,
2016, www.salon.com/2016/06/22/the_southwest_is_burning_partner/ (accessed November 23, 2019).

88 *"'a salt shaker of madness'"*: Elisha Fieldstadt, Hanna Guerrero, and Alex
Johnson, "Intense Heatwave Kills Four, Feeds Southwest Wildfires," NBC
News.com, June 19, 2016, www.nbcnews.com/news/weather/crews-fighting
-southwest-wildfires-prepare-excessive-heat-n595201 (accessed November 23, 2019).

88 *"progressively earlier snowmelt"*: No author listed, "Is Global Warming Fu-
eling Increased Wildfire Risks?," Union of Concerned Scientists, September
9, 2011, www.ucsusa.org/resources/global-warming-fueling-increased-wild
fire-risks (accessed November 23, 2019).

88 *"twelve thousand gallons of bright red chemicals"*: Fernanda Santos, "Raging
Wildfires in the Southwest Stretch Resources," *The New York Times*, June
22, 2016, www.nytimes.com/2016/06/23/us/raging-wildfires-in-the-southwest
-stretch-resources.html (accessed November 23, 2019).

88 *"many years made by Monsanto"*: Kira Hoffman, "It's Raining Red: Why
Use of Chemical Fire Retardants Is on the Rise," Science Borealis, February
12, 2018, blog.scienceborealis.ca/its-raining-red-why-use-of-chemical-fire
-retardants-is-on-the-rise/ (accessed November 23, 2019).

88 *"well-documented toxicity to aquatic life"*: Kostas D. Kalabokidis, "Effects
of Wildfire Suppression Chemicals on People and the Environment: A Review," *Global Nest: the International Journal* 2 (2000): 134–35.

88 *"disputed overall effectiveness"*: Eve Byron, "Effects of Fire Retardant on
Environment Assessed," *Independent Record*, May 15, 2011, helenair.com
/news/state-and-regional/effects-of-fire-retardant-on-environment-assessed
/article_5e3f41e0-7eb2-11e0-9cdb-001cc4c03286.html (accessed November
23, 2019).

88 *"41 million gallons of aerially delivered fire retardant"*: March 9, 2017, FOIA
request to Jennifer Jones, Forest Service.

88 *"pressure to use it often came from local communities"*: Lynda V. Mapes,
"Wildfire Retardant Flights Under Review; Some 'Just Painting Stuff Red,'"
The Seattle Times, September 26, 2015, www.seattletimes.com/seattle-news
/forest-service-reviewing-cost-effectiveness-of-aerial-retardants/ (accessed
November 23, 2019).

89 *"a massive wildfire charred over three million acres"*: Gerald W. Williams, *The USDA Forest Service—The First Century* (USDA Forest Service, 2005): 31, www.fs.fed.us/sites/default/files/media/2015/06/The_USDA_Forest_Service _TheFirstCentury.pdf (accessed November 23, 2019).

89 *"killed seventy-eight firefighters in a single afternoon"*: Michelle Nijhuis, "How Fire, Once a Friend of Forests, Became a Destroyer," *National Geographic*, November 22, 2015, www.nationalgeographic.com/news/2015/11/15 1122-wildfire-forest-service-firefighting-history-pyne-climate-ngbooktalk/ (accessed November 23, 2019).

89 *"the awesome ravages of nature"*: National Forests Foundation, "Blazing Battles: The 1910 Fire and Its Legacy," Your National Forests Magazine, www.nationalforests.org/our-forests/your-national-forests-magazine/blaz ing-battles-the-1910-fire-and-its-legacy (accessed November 23, 2019).

89 *"'the nearest thing there is to war'"*: Timothy Egan, *The Big Burn: Teddy Roosevelt and the Fire That Saved America* (Boston: Mariner Books, 2010), 152.

89 *"still disputing their toxicity"*: Judith Lewis Mernit, "Fire Fight: Forest Service Explores Chemical Retardant Hazards," *High Country News*, June 15, 2011, www.hcn.org/issues/43.10/fire-fight-forest-service-finally-reveals-the -hazards-of-chemical-retardants (accessed November 23, 2019); Byron, Eve, "Effects of Fire Retardant on Environment Assessed," *Independent Record*, May 15, 2011, helenair.com/news/state-and-regional/effects -of-fire-retardant-on-environment-assessed/article_5e3f41e0-7eb2-11e0 -9cdb-001cc4c03286.html (accessed November 23, 2019).

89 *"the efforts of a former timber lobbyist named Andy Stahl"*: John Cramer, "Retardant the New Frontier in Fire-Policy Fight," *Independent Record*, March 24, 2008, helenair.com/news/state-and-regional/retardant-the-new -frontier-in-fire-policy-fight/article_835b09d9-5ae6-5f85-87c2-e56ce3cbed d8.html (accessed November 23, 2019).

90 *"Forest Service agreed to expand its testing"*: Judith Lewis Mernit, "Fire Fight: Forest Service Explores Chemical Retardant Hazards," *High Country News*, June 15, 2011, www.hcn.org/issues/43.10/fire-fight-forest-service -finally-reveals-the-hazards-of-chemical-retardants (accessed November 23, 2019).

90 *"contrary findings from the Fish & Wildlife Service"*: Judith Lewis Mernit, "Fire Fight: Forest Service Explores Chemical Retardant Hazards," *High Country News*, June 15, 2011, www.hcn.org/issues/43.10/fire-fight-forest -service-finally-reveals-the-hazards-of-chemical-retardants (accesse No-

vember 23, 2019); Bruce Finley, "Wildfire: Red Slurry's Toxic Dark Side," *The Denver Post*, June 16, 2012, www.denverpost.com/2012/06/16/wildfire -red-slurrys-toxic-dark-side/ (accessed November 23, 2019).

90 *"new measures still being implemented"*: Judith Lewis Mernit, "Fire Fight: Forest Service Explores Chemical Retardant Hazards," *High Country News*, June 15, 2011, www.hcn.org/issues/43.10/fire-fight-forest-service-finally-re veals-the-hazards-of-chemical-retardants (accessed November 23, 2019).

90 *"50 percent of all aerial fire retardants were still deployed improperly"*: Lynda V. Mapes, "Wildfire Retardant Flights Under Review; Some 'Just Painting Stuff Red,'" *The Seattle Times*, September 26, 2015, www.seattle times.com/seattle-news/forest-service-reviewing-cost-effectiveness-of -aerial-retardants/ (accessed November 23, 2019).

90 *"fire retardants that suffused our mattresses, furniture, electronics"*: Connie Thompson, "Local Study Finds Toxic Flame Retardants in New TVs as CPSC Advances Toxin Ban," KomoNews, September 20, 2017, komonews .com/news/consumer/local-study-finds-toxic-flame-retardants-in-new-tvs -as-cpsc-advances-toxin-ban (accessed November 23, 2019).

90 *"tweak a molecule and reintroduce it"*: Smith and Lourie, *Slow Death by Rubber Duck: The Secret Danger of Everyday Things*, 121; David Brancac-cio, Daniel Shin, and Redmond Carolipio, "Examining the Toxic History of Flame Retardants," Marketplace, August 17, 2018, www.marketplace .org/2018/08/17/business/examining-toxic-history-flame-retardants (accessed November 23, 2019).

90 *"fire retardants were found in 100 percent of dust samples"*: Jamie Lincoln Kitman, "Word Than Lead?," *The Nation*, August 15, 2018, www.thenation .com/article/worse-than-lead/ (accessed November 23, 2019).

CHAPTER 10: *The Story of Color*

93 *"cotton production in the United States increased"*: Henry Hillard Earl, *Cen-tennial History of Fall River, Massachusetts* (New York: Atlantic Publishing and Engraving Company, 1877): 72.

93 *"fashion magazines featuring haute couture"*: Temma Balducci and Heather Belnap, eds., "Representing the Modern Woman: The Fashion Plate Recon-sidered (1865–75)," In *Women, Femininity and Public Space in European Visual Culture, 1789–1914*, 97–114 (London: Routledge, 2014), 105.

93 *"the elites sought to redraw the class lines"*: Temma Balducci and Heather Belnap, eds., "Representing the Modern Woman: The Fashion Plate

Reconsidered (1865–75)," In *Women, Femininity and Public Space in European Visual Culture, 1789–1914,* 97–114 (London: Routledge, 2014), 98.

93 *"dyes that required painstaking labor"*: Anthony S. Travis, *The Rainbow Makers: The Origins of the Synthetic Dyestuffs Industry in Western Europe* (Bethlehem: Lehigh University Press, 1993), 1.

93 *"bred in Mexican cacti"*: H. Losada et al., "The Presence and Experimental Utilisation of the 'Nopal' Vegetable (*Opuntia ficus-indica*) as an Important Sustainable Crop of Terraced Areas," *Livestock Research for Rural Development* 8, no. 2 (July 1996).

93 *"bat guano imported from Peru"*: Travis, *The Rainbow Makers,* 43.

93 *"a plant called woad"*: Travis, *The Rainbow Makers,* 1.

93 *"the cloth wrappings of mummies"*: J. Sérgio Seixas de Melo et al., "Spectral and Photophysical Studies of Substituted Indigo Derivatives in Their Keto Forms," *ChemPhysChem* 7, no. 11 (November 2006): 2303.

94 *"Perkin's grandfather was an alchemist"*: Simon Garfield, *Mauve: How One Man Invented a Color That Changed the World* (London: Faber and Faber, 2000), 16.

94 *"considered forming a touring string quartet"*: Travis, *The Rainbow Makers,* 34; Garfield, *Mauve,* 18.

94 *"trudging the London and Blackwall railway"*: Garfield, *Mauve,* 36.

94 *"celebrated talks at the Royal Institution"*: Garfield, *Mauve,* 18.

95 *"little practical application for chemistry"*: Garfield, *Mauve,* 29.

95 *"August Wilhelm von Hofmann . . . enlisted a mix of industry figures"*: Garfield, *Mauve,* 20.

95 *"grew almost exclusively in Bolivia and Peru"*: Garfield, *Mauve,* 32.

95 *"their colonial plantations in Asia"*: Benjamin Lewis Rice, *Mysore: A Gazetteer Compiled for Government,* Vol. 1 (Westminster: A. Constable, 1897), 892.

95 *"The laboratory was rudimentary"*: Robert Chenciner, *Madder Red: A History of Luxury and Trade* (London: Routledge, 2000), 260.

95 *"a rich, fade-resistant purple"*: Travis, *The Rainbow Makers,* 36.

96 *"purple dyes were made from lichens"*: Gavin Weightman, *The Industrial Revolutionaries: The Making of the Modern World, 1776–1914* (New York: Grove Press, 2010), 377.

96 *"the glandular mucus of snails"*: I. Irving Ziderman, "Purple Dyes Made from Shellfish in Antiquity," *Review of Progress in Coloration* 16, (June 1986): 46.

96 *"lost with the fall of Constantinople in 1453"*: Chenciner, *Madder Red,* 293.

96 *"the unusual shade of its drool"*: Garfield, *Mauve,* 39.

96 *"twelve thousand snails were required"*: Garfield, *Mauve,* 39.

96 *"the stench of the dyeing vats was ghastly"*: Jacqueline Banerjee, "Sir William Henry Perkin and the Coal-Tar Colours," The Victorian Web, www.victorian web.org/science/perkin.html (accessed November 23, 2019).

96 *"the purple was an irrelevant accident"*: Garfield, *Mauve*, 44.

96 *"Eugénie of France . . . decided that the color purple complemented her eyes"*: sdc.org.uk/about-us/perkin-legacy/mauve-150-years-of-colour/ (accessed November 23, 2019).

96 *"many of which featured hand-colored plates"*: Temma Balducci and Heather Belnap, eds., "Representing the Modern Woman: The Fashion Plate Reconsidered (1865–75)," In *Women, Femininity and Public Space in European Visual Culture, 1789–1914*, 97–114 (London: Routledge, 2014), 105.

96 *"Queen Victoria wore purple"*: Garfield, *Mauve*, 61.

96 *"'(. . . a rage for your color has set in . . .)'"*: Travis, *The Rainbow Makers*, 45

97 *"the streets of London and Paris were awash"*: Travis, *The Rainbow Makers*, 84.

97 *"Nicholson's blue, Martius yellow, and Hofmann's violet"*: Garfield, *Mauve*, 69.

97 *"'Language, indeed, fails adequately to describe the beauty'"*: John Joseph Beer, *The Emergence of the German Dye Industry* (Chicago: University of Illinois Press, 1959), 30.

98 *"the Germans did by developing the first industrial research laboratories"*: Graham D. Taylor and Patricia E. Sudnik, *Du Pont and the International Chemical Industry* (Boston: Twayne Press, 1984), 10.

98 *"embracing a more rigorous, theory-based methodology"*: Travis, *The Rainbow Makers*, 163.

98 *"valuable raw materials for future products"*: Taylor and Sudnik, *Du Pont and the International Chemical Industry*, 10.

98 *"bore a strong chemical resemblance to dyes"*: Beer, *The Emergence of the German Dye Industry*, 97.

98 *"synthesizing indigo, introduced aspirin"*: Steen, *The American Synthetic Organic Chemicals Industry*, 26.

98 *"followed by a laxative two years later"*: Steen, *The American Synthetic Organic Chemicals Industry*, 30.

98–99 *"Hermann Wilhelm Vogel noticed that coal tar colors"*: Garfield, *Mauve*, 10; H. Vogel, "On the Sensitiveness of Bromide of Silver to the So-Called Chemically Inactive Colours," *Chemical News*, December 26, 1873: 318–19, copying from *The Photographic News*, date and page not cited but apparently December 12, 1873, in turn translated from Vogel's own publication *Photographische Mittheilungen* 10, no. 117 (December 1873): 233–37.

99 *"Leo Baekeland, invented Bakelite, the earliest form of plastic"*: Steen, *The*

American Synthetic Organic Chemicals Industry, 282–83; Beer, *The Emergence of the German Dye Industry*, 100.

CHAPTER 11: *Poison Unleashed*

108 *"his father, Siegfried, never quite forgave him for it"*: Daniel Charles, *Master Mind: The Rise and Fall of Fritz Haber, the Nobel Laureate Who Launched the Age of Chemical Warfare* (New York: Ecco, 2005), 4.

108 *"A dye and pharmaceuticals salesman by trade"*: Diana Preston, *A Higher Form of Killing: The Secret History of Chemical and Biological Warfare* (New York: Random House, 2002), 78; Dietrich Stoltzenberg, *Fritz Haber: Chemist, Nobel Laureate, German, Jew* (Philadelphia: Chemical Heritage Foundation, 2005), 11.

108 *"'abandon, at all costs, the harbor . . .'"*: Charles, *Master Mind*, 6.

109 *"Haber got himself baptized"*: Charles, *Master Mind*, 30.

109 *"'This otherwise so splendid man'"*: *The Collected Papers of Albert Einstein*, Anna Beck, trans., *Volume 5: The Swiss Years: Correspondence, 1902–1914* (English translation supplement) (Princeton: Princeton University Press, 1995): 366.

109 "Germany controlled around 90 percent": Taylor and Sudnik, *Du Pont and the International Chemical Industry*, 8.

109 *"(the drabness of British military uniforms . . .)"*: Garfield, *Mauve*, 146.

109 *"nitrogen . . . served as the basis for all modern explosives"*: Garfield, *Mauve*, 154.

109 *"the British blockaded German ports"*: Preston, *A Higher Form of Killing*, 82.

110 *"Haber supervised its first deployment"*: Preston, *A Higher Form of Killing*, 83.

110 *"a long line at a spot called Lover's Knoll"*: Preston, *A Higher Form of Killing*, 88.

110 *"red fezzes likely got their tint from German dyes"*: Preston, *A Higher Form of Killing*, 94.

110 *"rewarded Haber's boss with a bottle of pink champagne"*: Preston, *A Higher Form of Killing*, 112.

110 *"'bullets might still kill individual soldiers'"*: Charles, *Master Mind*, 173.

111 *"Haber didn't attend her funeral"*: Preston, *A Higher Form of Killing*, 112.

111 *"a cyanide-based pest control agent called Zyklon A"*: Preston, *A Higher Form of Killing*, 271.

111 *"reserved the right to take an additional 25 percent"*: Steen, *The American Synthetic Organic Chemicals Industry: War and Politics, 1910–1930* (Chapel Hill: University of North Carolina Press, 2014), 185.

111   *"as much to do with the industry's need to justify its continued existence as consumers' need"*: Steen, *American Synthetic Organic Chemicals Industry*, 111; Taylor and Sudnik, *Du Pont and the International Chemical Industry*, 59.

112   *"the automobile furnished a hundred new applications for chemicals"*: Steen, *American Synthetic Organic Chemicals Industry*, 17, 237, 282; Taylor and Sudnik, *Du Pont and the International Chemical Industry*, 60.

112   *"the chemical industry had switched from coal tar to petroleum"*: Steen, *American Synthetic Organic Chemicals Industry*, 17.

112   *"materials like neoprene and nylon"*: Taylor and Sudnik, *Du Pont and the International Chemical Industry*, 265.

112   *"'today a stylish woman can go to the theater'"*: Herman Mark, *Giant Molecules* (New York: Time Life Books, 1966), 16.

112   *"the use of food additives increased dramatically"*: Lauren S. Jackson, "Chemical Food Safety Issues in the United States: Past, Present, and Future," *Journal of Agricultural and Food Chemistry* 57, no. 18 (September 2009): 8162–8163; Ai Hisano, "Standardized Color in the Food Industry: The Co-Creation of the Food Coloring Business in the United States," 1870–1940. Harvard Business School General Management Unit Working Paper No. 17–037 (October 2016): 23.

112–113   *"he published* Human Ecology and Susceptibility to the Chemical Environment"*:* Matthew Smith, *Another Person's Poison: A History of Food Allergy* (New York: Columbia University Press, 2015): 97.

113   *"'the arsenic crowned Queen of the Ball'"*: Garfield, *Mauve*, 101.

113   *"A German chemist discovered the same thing eight years later"*: Garfield, *Mauve*, 101.

113   *"the German doctor Richard von Volkmann noticed an increase in scrotal cancer"*: Clarke Brian Blackadar, "Historical Review of the Causes of Cancer," *World Journal of Clinical Oncology* 7, no. 1 (February 2016): 55.

113   *"dismissing the call for a return to the organic dyes as impractical"*: Garfield, *Mauve*, 104.

113   *"Hugo Schweitzer, the head of the U.S. subsidiary of the Bayer Corporation"*: Charles C. Mann and Mark L. Plummer, *The Aspirin Wars: Money, Medicine, and 100 Years of Rampant Competition* (New York: Knopf, 1991): 38.

113   *"which led to the synthetic production of perfume"*: Garfield, *Mauve*, 10.

114   *"British forces had wired London demanding access"*: Charles, *Master Mind*, 164.

114   *"leaks into the landscape, polluting groundwater"*: Charles, *Master Mind*, 110.

114   *"Doctors could never"*: Charles, *Master Mind*, 203.

114 *"The force of the temptation'"*: Michael Faraday, *Lectures on Education, Delivered at the Royal Institution of Great Britain* (London: J. W. Parker & Son,
1854), 58. archive.org/details/oneducatlectures00royarich/page/58 (accessed
November 23, 2019).

CHAPTER 12: *Quackery's Great-Granddad*

117 *"'Looking out on this panorama of light'"*: Edward Abbey, *Desert Solitaire*
(New York: Ballantine Books, 1971), 267.

117 *"'a Zeo headband (to monitor brain waves and sleep cycles)'"*: Rebecca
Mead, "Better, Faster, Stronger," *The New Yorker*, September 5, 2011, www
.newyorker.com/magazine/2011/09/05/better-faster-stronger (accessed November 23, 2019).

117 *"'as if* The New England Journal of Medicine *had been hijacked by the editors
of the SkyMall catalog'"*: Dwight Garner, "New! Improved! Shape Up Your
Life!," *The New York Times*, January 7, 2011, www.nytimes.com/2011/01/07
/books/07book.html (accessed November 23, 2019).

118 *"His feats included running marathons"*: Wim Hof, interview with Tim Ferriss, The Tim Ferriss Show, October 29, 2015, www.youtube.com/watch?v=X
iQ7ka11QnQ (accessed November 23, 2019).

118 *"mastering these techniques enabled you to control the autonomic nervous
system"*: Pepij van Erp, "Wim Hof's Cold Trickery," January 1, 2016, www
.pepijnvanerp.nl/2016/01/wim-hof-method/ (accessed November 23, 2019).

119 *"first developed a taste for the cold in 1979"*: Scott Carney, "The Iceman
Cometh," ScottCarney.com. www.scottcarney.com/article/the-iceman-cometh/
(accessed November 23, 2019).

119 *"But it was the second jump"*: Cult of Frogs, "An Exclusive Interview: What
Advice Does Wim Hof Have for Shamans?," Cult of Frogs, December 2017,
thecultoffrogs.wordpress.com/tag/hof/ (accessed November 23, 2019).

119 *"'My method can give them back control'"*: Scott Carney, "The Iceman
Cometh," ScottCarney.com, www.scottcarney.com/article/the-iceman-cometh/
(accessed November 23, 2019).

119 *"He pounded beers in his podcast interviews"*: Wim Hof, interview with Joe
Rogan, *The Joe Rogan Experience*, October 20, 2015, www.youtube.com
/watch?v=Np0jGp6442A (accessed November 23, 2019).

120 *"a study published in the* Proceedings of the National Academy of Sciences": Matthijs Kox et al., "Voluntary Activation of the Sympathetic Nervous System and Attenuation of the Innate Immune Response in Humans,"

Proccedings of the National Academy of Sciences of the United States of America 111, no. 20 (May 2014): 7381-3.

120 *"'I felt like I could conquer the world'"*: Roberta Attanasio, "'Iceman' Wim Hof and the Flow Within: The Immune System Goes with It," The Global Fool, May 20, 2014, theglobalfool.com/iceman-wim-hof-and-the-flow-with in-the-immune-system-goes-with-it/ (accessed November 23, 2019).

121 *"initiated the rapprochement between Thomas Jefferson and John Adams"*: John M. Kloos, *A Sense of Deity: The Republican Spirituality of Dr. Benjamin Rush* (New York: Carlson Publishing, 1991), 11.

122 *"He stumped for Washington"*: David Freeman Hawke, *Benjamin Rush: Revolutionary Gadfly* (Indianapolis: Bobbs-Merrill, 1971), 177.

122 *"infamously betrayed Washington"*: Hawke, *Benjamin Rush*, 213.

122 *"theories of humoral 'balance' dating back to Hippocrates"*: Elaine G. Breslaw, *Lotions, Potions, Pills, and Magic: Health Care in Early America* (New York: New York University Press, 2014), 44.

122 *"bad air arising from swamps"*: Hawke, *Benjamin Rush*, 322.

122 *"'a pretty draught' of mercury"*: Hawke, *Benjamin Rush*, 27.

122 *"Doctors rarely touched their patients"*: Hawke, *Benjamin Rush*, 45.

122 *"Even when instruments . . . were eventually introduced"*: Breslaw, *Lotions, Potions, Pills, and Magic*, 47, 182.

122 *"Only a competent physician could determine"*: Breslaw, *Lotions, Potions, Pills, and Magic*, 48.

122 *"putrid fevers, nervous fevers, hospital fevers"*: Hawke, *Benjamin Rush*, 46.

123 *"By August the death toll was mounting"*: Benjamin Rush, *An Account of the Bilious Remitting Yellow Fever as it Appeared in the City of Philadelphia in the Year 1793* (Philadelphia: Thomas Dobson, 1794), 129.

123 *"Handshaking fell out of fashion"*: Mathew Carey, *A Short Account of the Malignant Fever Lately Prevalent in Philadelphia with a Statement of the Proceedings That Took Place on the Subject, in Different Parts of the United States* (Philadelphia: printed by the author, 1794), 22.

123 *"advising against ringing church bells"*: Rush, *An Account of the Bilious Remitting Yellow Fever*, 22.

123 *"believed the disease originated from a heap of rotting coffee"*: Rush, *An Account of the Bilious Remitting Yellow Fever*, 17.

123 *"believed the fever to be contagious"*: Hawke, *Benjamin Rush*, 101.

123 *"cigar smoking, chewing garlic"*: Carl Binger, *Revolutionary Doctor: Benjamin Rush, 1746–1813* (New York: Norton, 1966), 208.

123 *"Rush began by prescribing a mild course of treatment"*: Judith W. Leavitt

and Ronald L. Numbers, eds., "Do It Yourself the Sectarian Way," In *Sickness & Health in America: Readings in the History of Medicine and Public Health* (Madison: University of Wisconsin Press), 1997, 246.

123   *"The letter was from a Virginia physician"*: Saul Jarcho, "John Mitchell, Benjamin Rush, and Yellow Fever," *Bulletin of the History of Medicine* 31, no. 2 (March/April 1957): 134.

123   *"claimed to have achieved positive results"*: Rush, *An Account of the Bilious Remitting Yellow Fever*, 196.

124   *"The vomiting, one doctor wrote, was 'cyclonic'"*: William G. Rothstein, *American Physicians in the 19th Century: From Sects to Science* (Baltimore: Johns Hopkins University Press, 1992), 49.

124   *"An 'ill-timed* scrupulousness'"*: Rush, *An Account of the Bilious Remitting Yellow Fever*, 198.

124   *"typically prescribed 15 grains of jalap"*: Breslaw, *Lotions, Potions, Pills, and Magic*, 95.

124   *"(a toxic combination of mercury and chloride . . .)"*: Breslaw, *Lotions, Potions, Pills, and Magic*, 46.

124   *"two or three times the usual dose"*: Rothstein, *American Physicians in the 19th Century*, 50.

124   *"Enough, it was said, for a horse"*: Binger, *Revolutionary Doctor,* 117.

124   *"given every six hours"*: R. B. Sullivan, "Sanguine Practices: A Historical and Historiographic Reconsideration of Heroic Therapy in the Age of Rush," *Bulletin of the History of Medicine* 68, no. 2 (Summer 1994): 218.

124   *"sometimes three times a day"*: Sullivan, "Sanguine Practices," 218.

124   *"nearly a gallon in the space of five days"*: Binger, *Revolutionary Doctor,* 217.

124   *"Rush surely drained many of his patients dry"*: Binger, *Revolutionary Doctor,* 12, 228.

124   *"a 'potent quack,' a 'lunatic'"*: Binger, *Revolutionary Doctor,* 241.

124   *"a 'barefaced puff'"*: Binger, *Revolutionary Doctor,* 242.

124   *"'one of those great discoveries which are made'"*: David Armstrong and Elizabeth Metzger Armstrong, *The Great American Medicine Show* (Upper Saddle River: Prentice Hall, 1991), 7.

125   *"exsanguinating as many as 150 patients a day"*: Sullivan, "Sanguine Practices," 218.

125   *"Not even after four out of six of his own household"*: Breslaw, *Lotions, Potions, Pills, and Magic*, 95; Armstrong and Armstrong, *The Great American Medicine Show*, 6.

125   *"'All will end well'"*: Hawke, *Benjamin Rush*, 289.

125 *"killed 15 percent of Philadelphia's population"*: Judith W. Leavitt and Ronald L. Numbers, eds., "Politics, Parties, and Pestilence: Epidemic Yellow Fever in Philadelphia and the Rise of the First Party System," In *Sickness & Health in America: Readings in the History of Medicine and Public Health* (Madison: University of Wisconsin Press, 1997), 241.

125 *"'the most appalling collective disaster'"*: Sullivan, "Sanguine Practices," 218.

125 *"He fought valiantly for prison reform"*: Hawke, *Benjamin Rush*, 364.

125 *"He fought equally hard to reform the treatment of the mentally ill"*: Binger, *Revolutionary Doctor*, 255.

125 *"spoke out boldly against the institution of slavery"*: Hawke, *Benjamin Rush*, 104.

125 *"'it was the triumph of a principle'"*: Rush, *An Account of the Bilious Remitting Yellow Fever*, 204.

125 *"'The adventurous physician,' he wrote to a friend"*: Thomas Jefferson Randolph, ed., Letter from Thomas Jefferson to Dr. Wistar, June 21, 1807, In *Memoirs, Correspondence and Private Papers of Thomas Jefferson, Late President of the United States*, Vol. 4 (London: Henry Colburn and Richard Bently, 1829), 96.

126 *"a 'great man' was what Rush had always longed to become"*: James Thacher, *American Medical Biography, Or, Memoirs of Eminent Physicians Who Have Flourished in America* (Boston: Richardson & Lord and Cottons & Barnard, 1828), 56.

126 *"'Death presses upon him'"*: Gert H. Brieger, ed., "On the Causes of Death in Diseases That Are Not Incurable." In *Medical America in the Nineteenth Century* (Baltimore: Johns Hopkins Press, 1972), 96.

127 *"'heroic medicine'"*: Sullivan, "Sanguine Practices," 211.

127 *"took pride in remaining unmoved"*: Breslaw, *Lotions, Potions, Pills, and Magic*, 179.

127 *"brag about how much blood they took"*: Ira Rutkow, *Seeking the Cure: A History of Medicine in America* (New York: Scribner, 2010), 38.

127 *"Rush went on to teach heroic medicine"*: Breslaw, *Lotions, Potions, Pills, and Magic*, 141.

127 *"indoctrinating as many as three thousand medical students"*: Binger, *Revolutionary Doctor*, 263.

127 *"half of all college-trained doctors in the United States"*: Hawke, *Benjamin Rush*, ix.

127 *"one man established the standard of care for an entire country"*: Bernard A.

Weisberger, "The Paradoxical Doctor Benjamin Rush," *American Heritage* 27, no. 1 (December 1975). www.americanheritage.com/content/paradoxical -doctor-benjamin-rush (accessed November 23, 2019).

127 *"setting American medicine back . . . at least two generations"*: Breslaw, *Lotions, Potions, Pills, and Magic*, 99.

127 *"The practice became common among male midwives"*: Breslaw, *Lotions, Potions, Pills, and Magic*, 121.

127 *"his patients did not, and accordingly sought solutions elsewhere"*: Breslaw, *Lotions, Potions, Pills, and Magic*, 101.

127 *"('The doctors . . . gave her disease the name of galloping consumption . . .')"*: Samuel Thomson, *New Guide to Health, or Botanic Family Physician* (Boston: J. Q. Adams, 1835), 24.

127–128 *"nearly lost his wife to the bloody ministrations of a male midwife"*: Rothstein, *American Physicians in the 19th Century*, 130.

128 *"an old woman who knew all the local roots and herbs"*: Rothstein, *American Physicians in the 19th Century*, 129.

128 *"publishing a book that sold far and wide"*: Rothstein, *American Physicians in the 19th Century*, 131.

128 *"would not be out of place in a contemporary herbal supplement store"*: Rothstein, *American Physicians in the 19th Century*, 132.

128 *"'On the lab'rer's money Lawyers feast'"*: Armstrong and Armstrong, *The Great American Medicine Show*, 24.

128 *"Thomsonianism claimed three million followers"*: Armstrong and Armstrong, *The Great American Medicine Show*, 26.

128 *"two thirds of them were women"*: Rothstein, *American Physicians in the 19th Century*, 90.

128 *"homeopathy was the invention of a German doctor, Samuel Hahnemann"*: Rothstein, *American Physicians in the 19th Century*, 152.

128 *"a vanishingly small dose would suffice"*: Rothstein, *American Physicians in the 19th Century*, 155.

129 *"Introduced to the United States by a follower of Hahnemann"*: Arthur Wrobel, ed., "Washington Irving and Homeopathy," In *Pseudo-Science and Society in 19th-Century America* (Lexington: University Press of Kentucky, 1987), 168; Rothstein, *American Physicians in the 19th Century*, 158.

129 *"homeopathy began to pose a serious challenge"*: Wrobel, *Pseudo-Science and Society in 19th-Century America*, 4.

129 *"patients were won over by the un-heroicness of the doctors"*: Arthur Wrobel, ed., "Washington Irving and Homeopathy," In *Pseudo-Science and Society*

in 19th-Century America (Lexington: University Press of Kentucky, 1987), 168.

129   *"The first nationwide medical organization"*: Rutkow, *Seeking the Cure*, 48.

129   *"10 percent discounts to homeopathy patients"*: Armstrong and Armstrong, *The Great American Medicine Show*, 35.

129   *"Regular physicians . . . were being physically assaulted"*: Breslaw, *Lotions, Potions, Pills and Magic*, 174.

129   *"the appeal of homeopathy among the American aristocracy"*: Rothstein, *American Physicians in the 19th Century*, 160; John S. Haller Jr., *American Medicine in Transition: 1840–1910* (Chicago: University of Illinois Press, 1981), 109.

129   *"Nathaniel Hawthorne, Harriet Beecher Stowe, and Louisa May Alcott"*: Armstrong and Armstrong, *The Great American Medicine Show*, 34.

129   *"women were the early adopters"*: C. C. Hildreth, "On Letting Blood from the Jugular in the Diseases of Children," *American Journal of Medical Sciences* 13, no. 26 (April 1847): 369–74; "On the effects of Bloodletting on the Young Subject," *New York Journal of Medicine and Collateral Sciences* 10, no. 30 (May 1848): 308; Rothstein, *American Physicians in the 19th Century*, 53.

130   *"kicked in the face by a horse"*: Arthur Wrobel, "Hydropathy, or the Water-Cure," In *Pseudo-Science and Society in 19th-Century America*, 74. Lexington: University Press of Kentucky, 1987.

130   *"(Sylvester Graham considered sperm a precious liquid . . .)"*: Ira Rutkow, *Seeking the Cure: A History of Medicine in America* (New York: Scribner, 2010), 47.

130   *"the importance of eating fresh food"*: Rothstein, *American Physicians in the 19th Century*, 133.

130   *"'an experimentally derived pharmacopoeia'"*: Rothstein, *American Physicians in the 19th Century*, 31.

130   *"spending sufficient time with patients to get a full picture"*: Breslaw, *Lotions, Potions, Pills, and Magic*, 159.

130   *"the importance of exercise and a comfortable convalescence"*: Arthur Wrobel, "Robert H. Collyer's Technology of the Soul," In *Pseudo-Science and Society in 19th-Century America* (Lexington: University Press of Kentucky, 1987), 31.

130   *"offering gentle cures for their children and manuals"*: Arthur Wrobel, "Hydropathy, or the Water-Cure," In *Pseudo-Science and Society in 19th-Century America* (Lexington: University Press of Kentucky, 1987), 92.

130 *"one of the founding fathers of vegetarianism"*: J. C. Whorton, "Historical Development of Vegetarianism," *The American Journal of Clinical Nutrition* 59, no. 5 (May 1994): 1104.

131 *"Thus he advised his students to always listen to old wives' tales"*: Hawke, *Benjamin Rush: Revolutionary Gadfly*, 378; Wayland D. Hand, ed., "The Interrelationship of Scientific and Folk Medicine in the United States of America Since 1850," In *American Folk Medicine* (Berkeley: University of California Press, 1980), 90.

131 *"'from the inventions and temerity of quacks'"*: Benjamin Rush, *Medical Inquiries and Observations* (Philadelphia: Prichard & Hall, 1789), 168.

132 *"the insights of Charcot, Breuer, Janet, and Freud"*: Rothstein, *American Physicians in the 19th Century*, 211.

132 *"phrenology . . . set the stage for neurology"*: Rothstein, *American Physicians in the 19th Century*, 28.

CHAPTER 13: *Toughing It Out*

134 *"'it's exactly what a zombie apocalypse would smell like'"*: Forum post: "Where in Flagstaff can you smell the Purina dog food plant?," Reddit.com, June 2, 2014, www.reddit.com/r/Flagstaff/comments/274mql/where_in_flag staff_can_you_smell_the_purina_dog/ (accessed November 23, 2019).

134 *"forced Purina to study how the stench might be mitigated"*: Dave Zorn, "Nestle-Purina Plans to Reduce Odor in Flagstaff over a Four-Year Period," KAFF News, June 1, 2016, gcmaz.com/nestle-purina-plans-to-reduce-odor-in-flagstaff-over-a-four-year-period/ (accessed November 23, 2019).

134 *"unacceptably high levels of a variety of other contaminants"*: Clean Label Project, "Pet Food Project," 2017, www.cleanlabelproject.org/pet-food/ (accessed November 23, 2019).

134 *"BPA, cadmium, even mycotoxins"*: Susan Thixton, "Double Dose of Pet Food Toxins," Truthaboutpetfood.com, April 13, 2018, truthaboutpetfood .com/double-dose-of-pet-food-toxins/ (accessed November 23, 2019).

134 *"euthanized cattle were used as ingredients"*: Robert Glatter, M.D., "Potentially Lethal Drug Found in Popular Dog Food Brands: What You Need to Know," *Forbes*, February 19, 2018, www.forbes.com/sites/robertglatter /2018/02/19/potentially-lethal-drug-found-in-popular-dog-food-brands -what-you-need-to-know/#14d05b52838e (accessed November 23, 2019).

134 *"'the finest natural ingredients'"*: bluebuffalo.com (accessed November 23, 2019).

134 *"'egg shell, raw feather and leg scale'"*: Susan Thixton, "Double Dose of Pet Food Toxins," Truthaboutpetfood.com, May 9, 2015. truthaboutpet food.com/purina-vs-blue-buffalo-fight-continues (accessed November 23, 2019); Tim Wall, "Purina and Blue Buffalo Settle Lawsuits After Two Years," PetfoodIndustry.com, November 3, 2016, www.petfoodindustry.com /articles/6113-purina-and-blue-buffalo-settle-lawsuits-after-two-years (accessed November 23, 2019).

135 *"(A close look at the ingredients of Chef Boyardee ravioli . . .)"*: Kristen Michaelis, "Decoding Labels: Chef Boyardee Beef Ravioli," Food Renegade, September 18, 2013, www.foodrenegade.com/decoding-labels-chef-boyardee -beef-ravioli/ (accessed November 23, 2019).

136 *"nothing more than a baroque form of PTSD"*: Herman Staudenmayer, "Post-Traumatic Stress Syndrome (PTSS): Escape in the Environment," *Journal of Clinical Psychology* 43, no. 1 (January 1987): 156.

136 *"much as the Academy of Medicine had dismissed Gulf War Illness"*: Deborah Cory-Slechta and Roberta Wedge, eds., *Gulf War and Health: Volume 10: Update of Health Effects of Serving in the Gulf War, 2016* (Washington, D.C.: National Academies Press, 2016), 240.

136 *"a biochemical connection between the two phenomena"*: Martin L. Pall, "Common Etiology of Posttraumatic Stress Disorder, Fibromyalgia, Chronic Fatigue Syndrome and Multiple Chemical Sensitivity via Elevated Nitric Oxide/Peroxynitrite," *Medical Hypotheses* 57, no. 2 (August 2001): 140.

136 *"link between early-life stress and the likelihood of a dysregulated immune response"*: J. B. Rayman et al., "Genetic Perturbation of TIA1 Reveals a Physiological Role in Fear Memory," *Cell Reports* 26, no. 11 (March 12, 2019): 2970; W. Wang et al., "Characteristics of Pro- and Anti-Inflammatory Cytokines Alteration in PTSD Patients Exposed to a Deadly Earthquake," *Journal of Affective Disorders* 248 (April 1, 2019): 53; P. F. Kuan et al., "Cell Type-Specific Gene Expression Patterns Associated with PostTraumatic Stress Disorder in World Trade Center Responders," *Translational Psychiatry* 9, no. 1 (January 15, 2019): 7; H. Hori and Y. Kim, "Inflammation and Post-Traumatic Stress Disorder," *Psychiatry and Clinical Neurosciences* 73, no. 4 (April 2019): 144; A. Shalev, "Long-Term Immune Alterations Accompanying Chronic Posttraumatic Stress Disorder Following Exposure to Suicide Bomb Terror Incidents during Childhood," *Neuropsychobiology* 76, no. 3 (2017): 2; H. Song, "Association of Stress-Related Disorders with Subsequent Autoimmune Disease," *JAMA* 319, no. 23 (June 19, 2018): 2388.

136 *"(it was only the Vietnam War that gave PTSD clinical legitimacy . . .)"*: David J. Morris, *The Evil Hours: A Biography of Post-Traumatic Stress Disorder* (New York: Mariner Books, 2016), 17.

136 *"why EI disproportionately affected women"*: I. R. Bell et al., "Sensitization to Early Life Stress and Response to Chemical Odors in Older Adults," *Biological Psychiatry* 35, no. 11 (June 1994): 857; Herman Staudenmayer, M. E. Selner, and J. C. Selner, "Adult Sequelae of Childhood Abuse Presenting as Environmental Illness," *Annals of Allergy* 71, no. 6 (December 1993): 538; Dava Stewart, "Major Depressive Disorder May Be Linked to Early Childhood Stress," MDMag.com, October 11, 2017, www.mdmag.com/medical-news /major-depressive-disorder-may-be-linked-to-early-childhood-stress (accessed November 23, 2019).

138 *"air pollution, which caused 6.5 million premature deaths every year"*: Stanley Reed, "Study Links 6.5 Million Deaths Each Year to Air Pollution," *The New York Times*, June 26, 2016, www.nytimes.com/2016/06/27/business/energy -environment/study-links-6-5-million-deaths-each-year-to-air-pollution .html (accessed November 23, 2019).

138 *"biggest emitters of airborne mercury and arsenic"*: NIH, US National Library of Medicine, ToxTown, "Power Plants," toxtown.nlm.nih.gov/sources -of-exposure/power-plants (accessed November 23, 2019).

138 *"one seventieth of a teaspoon of mercury"*: Union of Concerned Scientists, "Coal and Air Pollution," UCSUSA.org, December 19, 2017, www.ucsusa .org/clean-energy/coal-and-other-fossil-fuels/coal-air-pollution (accessed November 23, 2019).

138 *"not quite so negligible as the company claimed"*: Leslie Kaufman, "Mutated Trout Raise New Concerns Near Mine Sites," *The New York Times*, February 22, 2012, www.nytimes.com/2012/02/23/science/earth/mutated -trout-raise-new-concerns-over-selenium.html (accessed November 23, 2019); Rocky Barker, "The Full Story Behind Simplot's Two-Headed Fish and Phosphate Mine," *Idaho Statesman*, June 17, 2012, www.idahostatesman .com/2012/06/17/2158158/understanding-simplots-mutant.html via archive .is (accessed November 23, 2019).

139 *"fixed attribute of the individual"*: Frank Furedi, "The Only Thing We Have to Fear Is the 'Culture of Fear' Itself," Spiked, April 4, 2007, www.spiked-on line.com/2007/04/04/the-only-thing-we-have-to-fear-is-the-culture-of-fear -itself/ (accessed November 23, 2019).

CHAPTER 14: *Hyenas, Nukes, and Tornadoes*

140 *"Ulrich Beck, who . . . coined the term 'Risk Society'"*: Stuart Jeffries, "Risky Business: An Interview with Ulrich Beck," *The Guardian*, February 11, 2006, www.theguardian.com/books/2006/feb/11/society.politics (accessed November 23, 2019).

140 *"the production and distribution of bads"*: Ulrich Beck, interview with anonymous interviewer, ShortCutsTV, August 30, 2018, www.youtube.com /watch?v=BrP7m7BH-0o (accessed November 23, 2019).

140 *"'The injured of Chernobyl'"*: Ulrich Beck, "World Risk Society as Cosmopolitan Society? Ecological Questions in a Framework of Manufactured Uncertainties," *Theory, Culture & Society* 13, no. 4 (November 1, 1996): 31.

140 *"couldn't even safeguard against arsenic"*: Jesse Hirsch, "Heavy Metals in Baby Food: What You Need to Know," *Consumer Reports*, August 16, 2018, www.consumerreports.org/food-safety/heavy-metals-in-baby-food/ (accessed November 23, 2019).

141 *"'logic of institutionalized non-coping'"*: Ulrich Beck, *World at Risk* (Cambridge: Polity Press, 2009), 27–29.

141 *"over $43 billion in insured losses"*: Press Release: "Terrorism and Insurance: 13 Years After 9/11 the Threat of Terrorist Attack Remains Real," Insurance Information Institute, September 9, 2004, www.iii.org/press-release /terrorism-and-insurance-13-years-after-9-11-the-threat-of-terrorist-attack -remains-real-090914 (accessed November 23, 2019).

141 *"government coverage was capped at $100 billion"*: No author listed, "Background on: Terrorism Risk and Insurance," Insurance Information Institute, September 9, 2019, www.iii.org/article/background-on-terrorism -risk-and-insurance (accessed November 23, 2019).

141 *"(. . . the deaths of half a million people)"*: Matthew Bunn, Anthony Wier, and John Holdren, *Controlling Nuclear Warheads and Materials: A Report Card and Action Plan* (Washington, D.C.: Nuclear Threat Initiative and the Project on Managing the Atom, Harvard University, March 2003): ix. www.nti .org/media/pdfs/controlling-nuclear-warheads-and-materials-2003.pdf (accessed November 23, 2019).

141 *"a 20 percent likelihood over the next ten years"*: Richrd G. Lugar, "The Lugar Survey on Proliferation Threats and Responses," United States Senator Richard G. Lugar, Chairman, Senate Foreign Relations Committee, June 2005: 6, fas.org/irp/threat/lugar_survey.pdf (accessed November 23, 2019).

142 *"ten bales of woolens from Pisa to Sicily"*: Humbert O. Nelli, "The Earliest

Insurance Contract. A New Discovery," *The Journal of Risk and Insurance* 39, no. 2 (June 1972): 215–16.

142 *"a derivation of the Italian* risico": Gaetano Liuzzo et al., "The Term Risk: Etymology, Legal Definition and Various Traits," *Italian Journal of Food Safety* 3 (2014): 36.

142 *"a legitimate business with London as its epicenter"*: Claudia Klüppelberg, Daniel Straub, and Isabell M. Welpe, eds., "Risk in Historical Perspective: Concepts, Contexts, and Conjunctions," In *Risk: A Multidisciplinary Introduction* (New York: Springer International Publishing, 2014), 3–35.

142 *"one of the modern world's insurance giants"*: Solomon Huebner, "The Development and Present Status of Marine Insurance in the United States," *The Annals of the American Academy of Political and Social Science* 26, Insurance (September 1905): 241–72.

142 *"appetite for security that came with the rise of the bourgeoisie"*: Claudia Klüppelberg, Daniel Straub, and Isabell M. Welpe, eds., "Risk in Historical Perspective: Concepts, Contexts, and Conjunctions," In *Risk: A Multidisciplinary Introduction* (New York: Springer International Publishing, 2014), 3–35.

142 *"as risk was gradually transformed into a commodity"*: Claudia Klüppelberg, Daniel Straub, and Isabell M. Welpe, eds., "Risk in Historical Perspective: Concepts, Contexts, and Conjunctions," In *Risk: A Multidisciplinary Introduction* (New York: Springer International Publishing, 2014), 3–35.

142 *"expertise based on personal experience and transmitted hierarchically"*: Arwen P. Mohun, *Risk: Negotiating Safety in American Society* (Baltimore: Johns Hopkins University Press, 2013): 1.

143 *"eventually gave rise to engine inspectors"*: Claudia Klüppelberg, Daniel Straub, and Isabell M. Welpe, eds., "Risk in Historical Perspective: Concepts, Contexts, and Conjunctions," In *Risk: A Multidisciplinary Introduction* (New York: Springer International Publishing, 2014), 3–35.

143 *"world's first head-on train collision"*: Author not listed, "The First Multiple Fatality Train Wreck . . . Right Here in the Old Dominion," Disasterous History, May 31, 2013, disasteroushistory.blogspot.com/2013/05/the-first-multiple-fatality-train.html (accessed November 23, 2019).

143 *"killing three and maiming thirty"*: Mohun, *Risk*, 91.

143 *"In the decades that followed, safety engineers began to appear"*: Mohun, *Risk*, 102.

143 *"Railroad accidents . . . gave rise to state railroad commissions"*: Mohun, *Risk*, 102.

143 *"fever epidemics led to new municipal sanitation laws"*: Charles E. Rosenberg, *The Cholera Years: The United States in 1832, 1849, and 1866* (Chicago: University of Chicago Press, 1962): 33.

143 *"the first federal laws regulating technology"*: Claudia Klüppelberg, Daniel Straub, and Isabell M. Welpe, eds., "Risk in Historical Perspective: Concepts, Contexts, and Conjunctions," In *Risk: A Multidisciplinary Introduction* (New York: Springer International Publishing, 2014), 3–35.

143 *"the first state-run, state-mandated social insurance"*: No author listed, "Social Security History: Otto von Bismarck," www.ssa.gov/history/ottob.html (accessed November 23, 2019).

144 *"paving the way for probabilistic toxicology"*: Claudia Klüppelberg, Daniel Straub, and Isabell M. Welpe, eds., "Risk in Historical Perspective: Concepts, Contexts, and Conjunctions," In *Risk: A Multidisciplinary Introduction* (New York: Springer International Publishing, 2014), 3–35.

144 *"'Risk and its scientific control must remain in step'"*: Hans Marquardt, Siegfried G. Schäfer, Roger O. McClellan, and Frank Welsch, eds., "History of Toxicology," In *Toxicology* (San Diego: Academic Press, 1999), 12.

144 *"the growth of a new class of expert actuaries"*: Claudia Klüppelberg, Daniel Straub, and Isabell M. Welpe, eds., "Risk in Historical Perspective: Concepts, Contexts, and Conjunctions," In *Risk: A Multidisciplinary Introduction* (New York: Springer International Publishing, 2014), 3–35.

144 *"George Campbell began urging Bell Telephone"*: Claudia Klüppelberg, Daniel Straub, and Isabell M. Welpe, eds., "Risk in Historical Perspective: Concepts, Contexts, and Conjunctions," In *Risk: A Multidisciplinary Introduction* (New York: Springer International Publishing, 2014), 3–35.

144 *"specific thresholds called 'maximum allowable concentrations'"*: Soraya Boudia and Nathalie Jas, eds., "Introduction: Science and Politics in a Toxic World," In *Toxicants, Health and Regulation Since 1945* (London: Pickering and Chatto, 2013), 6.

144 *"'solely the dose determines that a thing is not a poison'"*: Joseph F. Borzelleca, "Paracelsus: Herald of Modern Toxicology," *Toxicological Sciences* 53, No. 1 (January 1, 2000): 2–4.

144 *"(. . . how Paracelsus saved a plague-stricken village . . .)"*: John G. Hargrave, "Paracelsus: German-Swiss physician," Encyclopaedia Britannica, Oct 30, 2019, www.britannica.com/biography/Paracelsus (accessed November 23, 2019).

144–145 *"the goal was the complete elimination of potential harm"*: Soraya Boudia and Nathalie Jas, eds., "Introduction. The Greatness and Misery of

Science in a Toxic World," In *Powerless Science? Science and Politics in a Toxic World* (New York: Berghan, 2014), 6.

145 *"MACs were embraced by industrial medicine"*: Claudia Klüppelberg, Daniel Straub, and Isabell M. Welpe, eds., "Risk in Historical Perspective: Concepts, Contexts, and Conjunctions," In *Risk: A Multidisciplinary Introduction* (New York: Springer International Publishing, 2014), 3–35.

145 *"the Food, Drug, and Cosmetic Act was passed"*: No author listed, "Milestones in U.S. Food and Drug Law History," January 1, 2018, www.fda.gov/about -fda/fdas-evolving-regulatory-powers/milestones-us-food-and-drug-law -history (accessed November 23, 2018).

145 *"began being used to regulate pollution in the air, soil, and water"*: Soraya Boudia and Nathalie Jas, eds., "Introduction: Science and Politics in a Toxic World," In *Toxicants, Health and Regulation Since 1945* (London: Pickering and Chatto, 2013), 6.

145 *"new benchmarks like 'permissible dosage'"*: Soraya Boudia and Nathalie Jas, eds., "Introduction. The Greatness and Misery of Science in a Toxic World," In *Powerless Science? Science and Politics in a Toxic World* (New York: Berghan, 2014), 10.

145 *"some of these substances were dangerous at any level"*: Jean-Baptiste Fressoz, "The Lessons of Disasters: A Historical Critique of Postmodern Optimism," Books & Ideas, May 27, 2011, booksandideas.net/The-Lessons -of-Disasters.html (accessed November 23, 2019).

145 *"asphalt, solvents, light bulb sockets"*: Soraya Boudia and Nathalie Jas, eds., "Introduction. The Greatness and Misery of Science in a Toxic World," In *Powerless Science? Science and Politics in a Toxic World* (New York: Berghan, 2014), 6.

145 *"the passage of the Delaney Amendment"*: Proctor, *The Cancer Wars*, 83.

145 *"along with 65,000 of your neighbors"*: The Manhattan Engineer District of the United States Army, "The Atomic Bombings of Hiroshima and Nagasaki," June 29, 1946, www.gutenberg.org/cache/epub/685/pg685-images.html (accessed November 23, 2019).

145 *"exposure should be 'as low as possible'"*: Proctor, *The Cancer Wars*, 158.

145 *"Nukes . . . had a marvelously clarifying effect on the way we thought about . . . risk"*: Claudia Klüppelberg, Daniel Straub, and Isabell M. Welpe, eds., "Risk in Historical Perspective: Concepts, Contexts, and Conjunctions," In *Risk: A Multidisciplinary Introduction* (New York: Springer International Publishing, 2014), 21.

145–146 *"the potential to trigger thermonuclear war"*: POGO staff, "POGO letter to DOE Secretary Bodman Regarding Serious Safety Problems at Pantex, a Nuclear Weapons Assembly Facility," Project on Government Oversight, December 12, 2006, www.pogo.org/letter/2006/12/pogo-letter-to-doe-secretary-bodman-regarding-serious-safety-problems-at-pantex-nuclear-weapons-assembly-facility/ (accessed November 23, 2019).

146 *"one of history's overlooked geniuses, H. A. Watson"*: Mohammad Sadegh Javadi, Azim Nobakht, and Ali Meskarbashee, "Fault Tree Analysis Approach in Reliability Assessment of Power System," *International Journal of Multidisciplinary Sciences and Engineering* 2, no. 6 (September 2011): 46.

146 *"discrete probabilities for each link"*: W. E. Vesely et al., *Fault Tree Handbook* (Washington, D.C.: U.S. Government Printing Office, 1981): II-1.

146 *"the 'worst-case scenario'"*: Search for "worst-case" scenario, Google Books Ngram Viewer, 1950–2005: books.google.com/ngrams/graph?content=worst-case+scenario&year_start=1950&year_end=2005&corpus=15&smoothing=3&share=&direct_url=t1%3B%2Cworst%20-%20case%20scenario%3B%2Cc0 (accessed November 23, 2019).

146 *"they would not fully cover damages resulting from a reactor meltdown"*: Soraya Boudia and Nathalie Jas, eds., "Introduction: Science and Politics in a Toxic World," In *Toxicants, Health and Regulation Since 1945* (London: Pickering and Chatto, 2013), 26.

146 *"the act is currently set to expire in 2025"*: Soraya Boudia and Nathalie Jas, eds., "Introduction: Science and Politics in a Toxic World," In *Toxicants, Health and Regulation Since 1945* (London: Pickering and Chatto, 2013), 20.

147 *"even when everything went right the risk still lingered"*: Soraya Boudia and Nathalie Jas, eds., "Managing Scientific and Political Uncertainty: Environmental Risk Assessment in a Historical Perspective." In *Powerless Science? Science and Politics in a Toxic World* (New York: Berghan, 2014), 101.

147 *"products were just as good as they needed to be"*: Adam Burgess, "Risk in History: A Thematic Review," unpublished, 2016, 11, www.academia.edu/28286953/Risk_in_History_A_Thematic_Review (accessed November 24, 2019).

147 *"one of the greatest inventions of the 1970s"*: Soraya Boudia and Nathalie Jas, eds., "Managing Scientific and Political Uncertainty: Environmental Risk Assessment in a Historical Perspective," In *Powerless Science? Science and Politics in a Toxic World* (New York: Berghan, 2014), 98.

147 *"risk 'is not something to be feared but to be embraced'"*: Proctor, *The Cancer Wars*, 96.

147 *"the release of the Rasmussen Report"*: Claudia Klüppelberg, Daniel Straub, and Isabell M. Welpe, eds., "Risk in Historical Perspective: Concepts,

Contexts, and Conjunctions," In *Risk: A Multidisciplinary Introduction* (New York: Springer International Publishing, 2014), 22.

147–148 *"six orders of magnitude less than dying in a car accident"*: U.S. Nuclear Regulatory Commission. "Reactor Safety Study: An Assessment of Accident Risks in U.S. Commercial Nuclear Power Plants." October 1975, 112.

148 *"three months before the meltdown at Three Mile Island"*: Claudia Klüppelberg, Daniel Straub, and Isabell M. Welpe, eds., "Risk in Historical Perspective: Concepts, Contexts, and Conjunctions." In *Risk: A Multidisciplinary Introduction* (New York: Springer International Publishing, 2014), 24.

148 *"the first professional organization dedicated to actuarial risk assessment"*: Claudia Klüppelberg, Daniel Straub, and Isabell M. Welpe, eds., "Risk in Historical Perspective: Concepts, Contexts, and Conjunctions." In *Risk: A Multidisciplinary Introduction* (New York: Springer International Publishing, 2014), 14.

148 *"an entire cadre of risk experts emerged"*: Claudia Klüppelberg, Daniel Straub, and Isabell M. Welpe, eds., "Risk in Historical Perspective: Concepts, Contexts, and Conjunctions." In *Risk: A Multidisciplinary Introduction* (New York: Springer International Publishing, 2014), 21–22.

148 *"a full-on, level 3 probability risk assessment"*: P. M. Roelofsen and J. Van der Steen, *Level 3 PSA Guidelines* (Petten: Netherlands Energy Research Foundation, July 1994).

148 *"more than one hundred person-years of analysis"*: M. R. Hayns, "The Evolution of Probabilistic Risk Assessment in the Nuclear Industry," *Process Safety and Environmental Protection* 77, no. 3 (May 1999): 140.

148 *"(PRA) became a part of the federal plant licensing process"*: Claudia Klüppelberg, Daniel Straub, and Isabell M. Welpe, eds., "Risk in Historical Perspective: Concepts, Contexts, and Conjunctions." In *Risk: A Multidisciplinary Introduction* (New York: Springer International Publishing, 2014), 24.

148 *"the burgeoning field's first journal"*: No author listed. "History." Society for Risk Analysis. www.sra.org/history (accessed November 24, 2019).

148 *"the number had quintupled"*: Soraya Boudia and Nathalie Jas, eds., "Managing Scientific and Political Uncertainty: Environmental Risk Assessment in a Historical Perspective." In *Powerless Science? Science and Politics in a Toxic World* (New York: Berghan, 2014), 102.

148 *"when it came to what constituted 'acceptable risk'"*: Ulrich Beck, *Risk Society: Towards a New Modernity* (London: Sage, 1992), 57–71; Ulrich Beck, *Ecological Politics in an Age of Risk* (Cambridge and Oxford: Polity Press, 1995), 111–18.

148 *"80 percent of industry scientists still believed in carcinogenic thresholds"*: Proctor, *The Cancer Wars*, 161.

148–149 *"assembled an advocacy group called the American Industrial Health Council"*: Soraya Boudia and Nathalie Jas, eds., "Managing Scientific and Political Uncertainty: Environmental Risk Assessment in a Historical Perspective," In *Powerless Science? Science and Politics in a Toxic World* (New York: Berghan, 2014), 95.

149 *"'alternative means to elaborate scientific judgments'"*: Soraya Boudia and Nathalie Jas, eds., "Managing Scientific and Political Uncertainty: Environmental Risk Assessment in a Historical Perspective," In *Powerless Science? Science and Politics in a Toxic World* (New York: Berghan, 2014), 104.

149 *"the National Research Council released what came to be known as 'The Red Book'"*: Soraya Boudia and Nathalie Jas, eds., "Managing Scientific and Political Uncertainty: Environmental Risk Assessment in a Historical Perspective," In *Powerless Science? Science and Politics in a Toxic World* (New York: Berghan, 2014), 95.

149 *"'unless the potential benefits . . . outweigh the potential costs'"*: Executive Order 12291, Section 2B, www.archives.gov/federal-register/codification /executive-order/12291.html (accessed November 24, 2019).

149 *"it was adopted as the new international standard"*: Soraya Boudia and Nathalie Jas, eds., "Managing Scientific and Political Uncertainty: Environmental Risk Assessment in a Historical Perspective," In *Powerless Science? Science and Politics in a Toxic World* (New York: Berghan, 2014), 106.

149 *"public safety advocates, naturally, felt otherwise"*: Cordner, *Toxic Safety*, 61–81.

149 *"the benchmark for the risk of chemical exposure was the Vitruvian man"*: Carl F. Cranor, *Legally Poisoned: How the Law Puts Us at Risk from Toxicants* (Cambridge: Harvard University Press, 2011), 201.

150 *"the latter always had the advantage"*: Cordner, *Toxic Safety*, 81; Proctor, *The Cancer Wars*, 83.

150 *"one side thought society was strong"*: Proctor, *The Cancer Wars*, 171.

150 *"risk management became a matter of 'personal responsibility'"*: Soraya Boudia and Nathalie Jas, eds., "Introduction: Science and Politics in a Toxic World," In *Toxicants, Health and Regulation Since 1945* (London: Pickering and Chatto, 2013), 15.

150–151 *"personal fallout shelters, 9-volt smoke detectors"*: No author listed, "A Brief History of Smoke Alarms," MySmokeAlarm.org: www.mysmoke alarm.org/history-of-smoke-alarms/ (accessed November 24, 2019).

151   *"retractable seat belts"*: Arwen P. Mohun, "Constructing the History of Risk. Foundations, Tools, and Reasons Why," *Historical Social Research* 41, no. 1 (155): 33.

151   *"school shooter panic buttons"*: Holmes, Oliver, "Panic Buttons for Mass Shootings Go on Sale in the US," *The Guardian*, March 28, 2019, www .theguardian.com/world/2019/mar/28/us-mass-shootings-panic-buttons-on -sale (accessed November 24, 2019).

151   *"'the last strongholds of vernacular risk culture'"*: Adam Burgess, "Risk in History: A Thematic Review," unpublished, 2016, 14, www.academia .edu/28286953/Risk_in_History_A_Thematic_Review (accessed November 24, 2019).

151   *"large numbers of scientific studies . . . could not be replicated"*: Francie Diep, "In the Latest Effort to Replicate Scientific Studies, Only 62 Percent Hold Up Under Scrutiny," *Pacific Standard*, August 27, 2018, psmag.com /news/in-the-latest-effort-to-replicate-scientific-studies-only-62-percent -hold-up-under-scrutiny (accessed November 24, 2019).

151   *"BPA substitutes were equally dangerous as BPA"*: Robert F. Service, "BPA Substitutes May Be Just as Bad as the Popular Consumer Plastic," *Science*, September 13, 2018, www.sciencemag.org/news/2018/09/bpa-substitutes -may-be-just-bad-popular-consumer-plastic (accessed November 24, 2019).

151   *"'living with ineradicable non-knowing'"*: Ulrich Beck, *Risk Society: Towards a New Modernity* (London: Sage, 1992), 115.

151–152   *"one step ahead of the risks that they themselves created"*: Jean-Baptiste Fressoz, "The Lessons of Disasters: A Historical Critique of Postmodern Optimism," Books & Ideas, May 27, 2011, booksandideas.net/The-Lessons -of-Disasters.html (accessed November 23, 2019).

152   *"'the next generation of nuclear power plants will be even more stable'"*: Eric P. Loewen, Ph.D., "To Understand Fukushima We Must Remember Our Past: The History of Probabilistic Risk Assessment of Severe Reactor Accidents: Address to Sociedad Nuclear Mexicana Conference," August 8, 2011, www .ans.org/about/officers/docs/8-aug-11_mexico_address_fs.pdf (accessed November 24, 2019).

CHAPTER 15: *Snowflake*

155   *"'no preservatives, no caffeine, and no evil sciency chemical concoctions'"*: IZZE Clementine Sparkling Juice, 8.4 oz, 24ct Product Description, www

.amazon.com/IZZE-Clementine-Sparkling-Juice-24ct/dp/B00MTBIT9M (accessed November 24, 2019).

157 *"Dursban . . . banned for indoor use"*: Jon R. Luoma, "The Ban That Wasn't," *Mother Jones*, September/October 2000, www.motherjones.com/politics /2000/09/ban-wasnt/ (accessed November 24, 2019).

157 *"(. . . reversed by Trump's pick to lead the EPA)"*: Eric Beech, "U.S. EPA Denies Petition to Ban Pesticide Chlorpyrifos," Reuters, March 29, 2017, www.reuters .com/article/us-usa-pesticide-epa/u-s-epa-denies-petition-to-ban-pesticide -chlorpyrifos-idUSKBN17039F (accessed November 24, 2019).

162 *"at $300 per acre the prices couldn't be beat"*: Zillow.com, April 24, 2018.

CHAPTER 16: *The Figure in the Static*

167 *"so fast and hot that it created its own weather"*: Michael Johnson, "Rodeo-Chediski Fire Underscored Need to Thin Forest," *White Mountain Independent*, June 16, 2017, www.wmicentral.com/news/apache_county/rodeo -chediski-fire-underscored-need-to-thin-forest/article_b86b09ae-b995 -555a-b3af-d072dc7e6e17.html (accessed November 24, 2019).

167 *"a record 86,000 gallons of fire retardant were dropped"*: Dennis Wagner, "Rodeo-Chediski Fire: Painful Memories Linger," *The Arizona Republic*, June 17, 2012, archive.azcentral.com/arizonarepublic/news/articles /2012/06/17/20120617rodeo-chediski-fire-painful-memories.html (accessed November 24, 2019).

167 *"covering some houses"*: Arizona Department of Health Services Office of Environmental Health, "Public Health Assessment Rodeo-Chediski Fire June 18, 2002–July 9, 2002 Navajo County, Arizona," July 15, 2003: 10, www.azdhs.gov/documents/preparedness/epidemiology-disease-control /environmental-toxicology/rodeo_chedeski_assmnt.pdf (accessed November 24, 2019).

167–168 *"The brand was Fire-Trol"*: Arizona Department of Health Services Office of Environmental Health, "Public Health Assessment Rodeo-Chediski Fire June 18, 2002–July 9, 2002 Navajo County, Arizona," July 15, 2003: 10, www .azdhs.gov/documents/preparedness/epidemiology-disease-control/environ mental-toxicology/rodeo_chedeski_assmnt.pdf (accessed November 24, 2019).

168 *"22 and 390 times higher than EPA standards"*: Aregai Tecle and Daniel Neary, "Water Quality Impacts of Forest Fires," *Pollution Effects and Control* 3, no. 2 (July 2015): 3.

172 *"wandering the streets in their bathrobes"*: Nancy Owen Lewis, *Chasing the Cure in New Mexico: Tuberculosis and the Quest for Health* (Santa Fe: Museum of New Mexico Press, 2016), 124.

172 *"shoved through the windows of Pullman cars"*: Barbara Bates, *Bargaining for Life: A Social History of Tuberculosis, 1876–1938* (Philadelphia: University of Pennsylvania Press, 1992), 140; Lewis, *Chasing the Cure in New Mexico*, 9, 157.

172 *"the tents reached nearly to the foothills"*: Billy M. Jones, *Health-Seekers in the Southwest, 1817–1900* (Norman: University of Oklahoma Press, 1967), 170, 186.

172 *"the leading cause of death in the U.S."*: University of Virginia Historical Exhibits, "Early Research and Treatment of Tuberculosis in the 19th Century," exhibits.hsl.virginia.edu/alav/tuberculosis/ (accessed November 24, 2019).

172 *"recovering from disease required rising above 'the fever line'"*: Jones, *Health-Seekers in the Southwest*, 18.

172 *"line could be found at precisely six thousand feet"*: Lewis, *Chasing the Cure in New Mexico*, 30.

172 *"geography . . . became, like a drug, prescribable"*: Cindy S. Aron, *Working At Play: A History of Vacations in the United States* (Oxford: Oxford University Press, 1999), 22.

173 *"they . . . began collecting climatological data"*: Bates, *Bargaining for Life*, 27–28.

173 *"drying out the wet lungs of consumptives"*: University of Virginia Historical Exhibits, "Dry Air: Arid Climates to Cure Tuberculosis," exhibits.hsl.virginia.edu/breath/dry-air/ (accessed November 24, 2019).

173 *"(many survived only to later succumb—in vast numbers—to tuberculosis)"*: Lewis, *Chasing the Cure in New Mexico*, 162.

173 *"fresh air, good food, and absolute rest"*: Jones, *Health-Seekers in the Southwest*, 180.

173 *"local boards of health were actively marketing it"*: Jones, *Health-Seekers in the Southwest*, 122, 150; Lewis, *Chasing the Cure in New Mexico*, 9.

173 *"'Come West and Live!'"*: Sheila M. Rothman, *Living in the Shadow of Death: Tuberculosis and the Social Experience of Illness in American History* (Baltimore: Johns Hopkins University Press, 1995), 148.

173 *"20–25 percent came in search of better health"*: Jones, *Health-Seekers in the Southwest*, viii.

173 *"a tubercular Atlanta dentist named Doc Holliday"*: Jones, *Health-Seekers in the Southwest*, 119.

173 *"the fateful fork in the path for her errant kid"*: Lewis, *Chasing the Cure in New Mexico*, 6.

173 *"In New Mexico alone, over 170 were built"*: Anya Grahn, "Historic Tuberculosis Sanitariums: Geography and Climate as a Cure," National Trust for Historic Preservation, November 6, 2016, savingplaces.org/stories/historic -tuberculosis-sanitariums-geography-and-climate-as-a-cure-2#.WFGmaa IrLb5 (accessed November 24, 2019).

173 *"Koch's discovery of the TB bacillus"*: Helen Bynum, *Spitting Blood: The History of Tuberculosis* (Oxford: Oxford University Press, 2012), 105.

174 *"like Charcot's hysteria before it"*: Elaine Showalter, *Hystories* (New York: Columbia University Press, 1998), 37.

174 *"EI was most often compared to neurasthenia"*: Donald J. McGraw, M.D., M.Ph., "Multiple Chemical Sensitivities—Modern Medical Conundrum or Old Story with a New Title?," *Journal of Occupational and Environmental Medicine* 53, no. 1 (January 2011): 103; E. Shorter, "Multiple Chemical Sensitivity: Pseudodisease in Historical Perspective," *Scandinavian Journal of Work, Environment and Health* 23, Supplement 3 (1997): 40.

174 *"George M. Beard provided a nonexhaustive list of eighty of them"*: George M. Beard, *American Nervousness: Its Causes and Consequences, a Supplement to Nervous Exhaustion (Neuresthenia)* (New York: G.P. Putnam's Sons, 1881), 8–9.

174 *"'today it means almost anything and with equal truth almost nothing'"*: David G. Schuster, *Neurasthenic Nation: America's Search for Health, Happiness and Comfort, 1869–1920* (New Brunswick: Rutgers University Press, 2011), 141.

174 *"he substituted 'nerve-force'"*: Beard, *American Nervousness*, 9.

174 *"(. . . Beard—and, later, Freud) had also at some point suffered from it"*: Schuster, *Neurasthenic Nation* 16; Tom Lutz, *American Nervousness, 1903: An Anecdotal History* (Ithaca: Cornell University Press, 1993), 169.

174 *"neurasthenics, like sensitives, were frequently dismissed as 'malingerers'"*: Campbell, "The Making of 'American': Race and Nation in Neurasthenic Discourse," 167; Schuster, *Neurasthenic Nation*, 4.

174 *"'Americanitis,' it was sometimes called"*: Campbell, "The Making of 'American': Race and Nation in Neurasthenic Discourse," 164.

174 *"'the mental activity of women'"*: Beard, *American Nervousness*, 96.

175 *"at the cost of a newfangled anxiety"*: Beard, *American Nervousness*, 103.

175 *"(he literally compared the human body to a light bulb)"*: Beard, *American Nervousness*, 96.

175 *"somaticized dissent from 12.8 percent of the population"*: Steinemann, "National Prevalence and Effects of Multiple Chemical Sensitivities," 152.

176 *"'created out of the same industrial paradigm'"*: Pamela Reed Gibson, Ph.D., "Understanding & Accommodating People with Multiple Chemical Sensitivity in Independent Living," Independent Living Research Utilization, January 1, 2012, www.ilru.org/understanding-accommodating-people-with-multiple-chemical-sensitivity (accessed November 24, 2019).

176 *"a delusion induced by sensationalizing news outlets"*: Shorter, "Multiple Chemical Sensitivity: Pseudodisease in Historical Perspective," 41; Laurence M. Binder and Keith A. Campbell, "Medically Unexplained Symptoms and Neuropsychological Assessment," *Journal of Clinical and Experimental Neuropsychology* 26, no. 3 (2004): 371, 374.

176 *"eleven times greater symptoms than controls"*: J. E. Broderick, E. Kaplan-Liss, and E. Bass E., "Experimental Induction of Psychogenic Illness in the Context of a Medical Event and Media Exposure," *American Journal of Disaster Medicine* 6, no. 3 (May/June 2011): 6.

177 *"gluten was only dangerous to the 1 percent of the population"*: Greetje J. Tack, et al., "The Spectrum of Celiac Disease: Epidemiology, Clinical Aspects and Treatment," *Nature Reviews: Gastroenterology and Hepatology* 7, no. 4 (April 2010): 204.

177 *"30 percent of the population was trying to avoid it"*: Elaine Watson, "30% of US Adults Trying to Cut Down on Gluten, Claims NPD Group," FoodNavigator-USA.com, March 8, 2013, www.foodnavigator-usa.com/Article/2013/03/08/30-of-US-adults-trying-to-cut-down-on-gluten-claims-NPD-Group (accessed November 25, 2019).

177 *"tiny pits began appearing in car windshields"*: Leslie P. Boss, "Epidemic Hysteria: A Review of the Published Literature," *Epidemiologic Reviews* 19, no. 2 (1997), 233–43; R. E. Bartholomew, "RE: 'Epidemic Hysteria: A Review of the Published Literature,'" *American Journal of Epidemiology* 151, no. 2 (2000); www.neatorama.com/2012/09/17/The-Great-Seattle-Windshield-Epidemic/ (accessed November 25, 2019).

177 *"To weed out the old and infirm"*: usatoday30.usatoday.com/weather/science/2001-03-07-contrails.htm (accessed November 25, 2019).

177 *"'Mass psychogenic illness,' a related term"*: Robert E. Bartholomew, *Little Green Men, Meowing Nuns and Head-Hunting Panics: A Study of Mass Psychogenic Illness and Social Delusion* (Jefferson: MacFarland & Co., 2001), 8–9; L. A. Page, "Frequency and Predictors of Mass Psychogenic Illness," *Epidemiology* 21, no. 5 (September 2010): 744.

177–178 "(*girls' private schools, commonly*)": Donna M. Goldstein and Kira Hall, "Mass Hysteria in Le Roy, New York: How Brain Experts Materialized Truth and Outscienced Environmental Inquiry," *American Ethnologist* 42, no. 4 (2015) 640–57.

178 *"a woman convinced she had suddenly become allergic to meat"*: Megan Molteni, "Oh, Lovely: The Tick That Gives People Meat Allergies Is Spreading," *Wired*, June 17, 2017, www.wired.com/story/lone-star-tick-that-gives-people-meat-allergies-may-be-spreading/ (accessed November 25, 2019).

178 *"Thomas Platts-Mills had been telling people the same sort of thing for twenty years"*: Jad Abumrad and Robert Krulwich, "Alpha Gal," Radiolab, October 27, 2016, www.wnycstudios.org/story/alpha-gal (accessed November 25, 2019).

CHAPTER 17: *The Gospel of Dr. Gray*

184 *"research findings had been made linking EI with eighteen different genes"*: A. D. Rouillard, et al., "The Harmonizome: A Collection of Processed Datasets Gathered to Serve and Mine Knowledge About Genes and Proteins," Database (Oxford), July 3, 2016, amp.pharm.mssm.edu/Harmonizome/gene_set/Multiple+Chemical+Sensitivity/HuGE+Navigator+Gene-Phenotype+Associations (accessed November 25, 2019).

184 *"including the MTHFR gene"*: V. Loria-Kohen, "Multiple Chemical Sensitivity: Genotypic Characterization, Nutritional Status and Quality of Life in 52 Patients," *Medicina Clinica* 149, no. 4 (August 22, 2017): 141–146.

184 *"these findings were frustratingly inconsistent"*: Loria-Kohen, "Multiple Chemical Sensitivity: Genotypic Characterization, Nutritional Status and Quality of Life in 52 Patients," 141–46; T. M. Dantoft, "Gene Expression Profiling in Persons with Multiple Chemical Sensitivity Before and After a Controlled N-Butanol Exposure Session," *BMJ Open* 7, no. 2 (February 22, 2017): 1; X. Cui, "Evaluation of Genetic Polymorphisms in Patients with Multiple Chemical Sensitivity," *PLoS One* 8, no. 8 (August 13, 2013): 7; N. D. Berg, "Genetic Susceptibility Factors for Multiple Chemical Sensitivity Revisited," *International Journal of Hygiene and Environmental Health* 213, no. 2 (March 2010): 139.

184 *"more convincing genetics work had been done with myalgic encephalomyelitis"*: W. C. De Vega et al., "Integration of DNA Methylation & Health Scores Identifies Subtypes in Myalgic Encephalomyelitis/Chronic Fatigue Syndrome," *Epigenomics* 10, no. 5 (May 2018): 539; C. B. Nguyen, "Whole

Blood Gene Expression in Adolescent Chronic Fatigue Syndrome: An Exploratory Cross-Sectional Study Suggesting Altered B Cell Differentiation and Survival," *Journal of Translational Medicine* 15, no. 1 (May 11, 2017): 15; S. Marshall-Gradisnik, "Single Nucleotide Polymorphisms and Genotypes of Transient Receptor Potential Ion Channel and Acetylcholine Receptor Genes from Isolated B Lymphocytes in Myalgic Encephalomyelitis/ Chronic Fatigue Syndrome Patients," *The Journal of International Medical Research* 44, no. 6 (December 2016): 1381; K. A. Schlauch, "Genome-Wide Association Analysis Identifies Genetic Variations in Subjects with Myalgic Encephalomyelitis/Chronic Fatigue Syndrome," *Translational Psychiatry* 6, no. 2 (February 9, 2016): 6.

187 *"failing to meet EPA dust emissions standards"*: Roy Ing, M.D., "Community Asbestos Exposure in Globe, Arizona," *Journal of Pediatrics* 99, no. 3 (September 1981): 409.

187 *"The other was owned by a guy named Jack Neal"*: Marilyn Taylor, "Babbitt Studies Globe Relocation Issue," *Arizona Republic*, December 21, 1979.

187 *"a fifty-five-unit trailer park he dubbed 'Mountain View Estates'"*: "Superfund Record of Decision: Mountain View/Globe Site, AZ," Report Number EPA.ROD.R09.83.003, United States Environmental Protection Agency, Office of Emergency and Remedial Response, June 1983.

187 *"the sewage at Mountain View Estates was backing up"*: "Superfund Record of Decision: Mountain View/Globe Site, AZ," Report Number EPA.ROD. R09.83.003, United States Environmental Protection Agency, Office of Emergency and Remedial Response, June 1983.

187–188 *"calling the State Health Department's actions 'criminal'"*: No author listed, "Evacuation of Globe Area Urged," *Arizona Republic*, December 18, 1979.

188 *"the governor of Arizona declared a state of emergency"*: "Superfund Record of Decision: Mountain View/Globe Site, AZ," Report Number EPA, ROD.R09.83.003. United States Environmental Protection Agency, Office of Emergency and Remedial Response, June 1983.

188 *"'We never promised anyone money to relocate'"*: Reid, Gail, "Asbestos Risk Still Present, Globe Gripes," *Arizona Republic,* March 30, 1981.

CHAPTER 18: *Prescribing the North Wind*

192 *"Shoemaker had once defended himself along these lines"*: Ritchie C. Shoemaker, *Desperation Medicine* (Baltimore: Gateway Press, 2001), 40.

193 *"Rea began his career as a thoracic and cardiovascular surgeon"*: Leif, "Sick of the World," *D Magazine*, January 2004, www.dmagazine.com/publications/d-magazine/2004/january/sick-of-the-world/ (accessed November 25, 2019).

193 *"Rea began developing symptoms of his own"*: Ibid.

193 *"he tested for allergens and then prescribed ongoing injections"*: No author listed, "LDA versus EHC-D Skin Testing and Immunotherapy," Environmental Health Center, Dallas, www.ehcd.com/lda-versus-ehc-d-allergy-skin-testing/ (accessed November 25, 2019).

193 *"The list of substances for which he tested was 259 items long"*: Transcript: "Oral Videotaped Deposition of William J. Rea, in the Matter of the Complaint Against before the Texas Medical Board," SOAH Docket no. 503-07-4032, May 21, 2010: 142-62.

193 *"everything from men's cologne to rug padding"*: Transcript: "Oral Videotaped Deposition of William J. Rea," 160.

194 *"he claimed to have treated over thirty thousand patients"*: Transcript: "Oral Videotaped Deposition of William J. Rea," 45.

194 *"his views were embraced by the aristocracy"*: Whitley, Glenna. "Is the 20th Century Making You Sick?," *D Magazine*, August 1990, www.dmagazine.com/publications/d-magazine/1990/august/is-the-20th-century-making-you-sick/ (accessed November 23, 2019).

194 *"Rea naturally took up his mantle"*: Staudenmayer, *Environmental Illness,* 92–97.

194 *"'a legend within our ranks'"*: Horner, Kim, "State Board Calls Methods of Doctor 'Pseudoscience,'" *Dallas Morning News*, October 30, 2007.

194 *"admitted that patients often spent $30,000"*: Lisa Nagy, interview by author, July 28, 2016.

194 *"he had been disciplined by the Texas Medical Board"*: Stephen Barrett, M.D., "Multiple Chemical Sensitivity: A Spurious Diagnosis," Quackwatch, www.quackwatch.org/01QuackeryRelatedTopics/mcs.html (accessed November 24, 2019).

194 *"ABC's* Nightline *covered the story"*: www.youtube.com/watch?v=-gx4zxxi0xQ (accessed November 25, 2019).

194 *"'there is currently an organized nation-wide effort to destroy the specialty'"*: William J. Rea, "Letter to Patients about Texas State Medical Board Disciplinary Action from Dr. William J. Rea," Environmental Health Association of Nova Scotia, November 18, 2007, www.environmentalhealth.ca/special/fall07reaopenletter.html (accessed November 25, 2019).

195   *"The tests he ran on his patients did 'not even qualify as experimental'"*: Transcript: "Oral Videotaped Deposition of William J. Rea," 579-81, 592-94.

195   *"he did not inject his patients with jet fuel"*: Transcript: "Oral Videotaped Deposition of William J. Rea," 115.

195   *"collected behind the chain link fence at the local airfield"*: Transcript: "Oral Videotaped Deposition of William J. Rea," 125.

195   *"the antigen he used contained no chemicals whatsoever"*: Transcript: "Oral Videotaped Deposition of William J. Rea," 131.

195   *"the north wind he collected came from right outside"*: Transcript: "Oral Videotaped Deposition of William J. Rea," 138.

196   *"he was also getting paid by Blue Cross/Blue Shield"*: Transcript: "Oral Videotaped Deposition of Keith E. Miller, M.D., F.A.A.F.P., Keith Miller vs. Shirley P. Pigott and Steven F. Hotze," Shelby County District Court, Texas, Cause No. 08-CV-29961, April 3, 2009: 64, 74; Louis Leichter, "Member of Texas Medical Board Resigns Amid Flood of Criticism," Texas Medical Licensing Law Blog, September 15, 2007 (accessed November 25, 2019).

196–197   *"the board had ignored the testimony of these five patients"*: Louis Leichter, "Member of Texas Medical Board Resigns Amid Flood of Criticism," Texas Medical Licensing Law Blog, September 15, 2007 (accessed November 25, 2019).

197   *"twice compared to the Third Reich"*: Transcript: "Texas Legislative Hearing, Subcommittee on Regulatory, in Oversight of the Texas Medical Board (TMB)," October 23, 2007: 203, 210.

200   *"Kalafut stomped out of the room in a fury"*: Craig Malisow, "Dr. Steven Hotze's Weird War Against the Texas Medical Board," *Houston Press*, March 26, 2014, www.houstonpress.com/news/dr-steven-hotzes-weird-war -against-the-texas-medical-board-6601127 (accessed November 25, 2019).

201   *"the executive director of the TMB resigned"*: No author listed, "Questionable tactics," *Houston Chronicle*, April 15, 2008, www.chron.com/neighborhood /article/Questionable-tactics-9355098.php (accessed November 25, 2019).

201   *"followed shortly thereafter by the president of the board"*: Association of American Physicians and Surgeons, Inc.: Press Release: "AAPS Applauds the Resignation of TMB President Robert Kalafut, Mari Robinson Should Resign Next," December 12, 2008, www.aapsonline.org/press/texas-medi cal-board-president-resigns-press-release-12-12-09.php (accessed November 25, 2019).

201   *"an order that he provide patients . . . with a revised informed consent form"*: Texas Medical Board Press Release:" Medical Board Disciplines 187

Doctors and Issues 88 Licenses, Thursday, September 02, 2010, www.tmb.state
.tx.us/dl/0BCB015A-59AC-7E7E-BCA9-E2B7AC8E80F6&usg=AOv
Vaw1oxfeaEYy4DAUx5ghgTOJo (accessed November 25, 2019).

201 *"legislature passed a law prohibiting anonymous complaints"*: No author
listed, "Texas Makes It Harder to Be Whistleblower: New Law Prohibits
Anonymous Complaints to the State Medical Board," Advisory Board, September 20, 2011, www.advisory.com/daily-briefing/2011/09/20/identifying
-whistleblowers (accessed November 25, 2019).

201 *"fight[ing] the government takeover of medicine'"*: Barry Meier, "Vocal
Physicians Group Renews Health Law Fight," *The New York Times*, January 18, 2011, www.nytimes.com/2011/01/19/business/19physicians.html (accessed November 25, 2019).

201 *"doctors who have been mugged by Medicare'"*: No author listed, "About
AAPS," Association of American Physicians and Surgeons, aapsonline.org
/about-aaps/ (accessed November 25, 2019).

201 *"more widely known for opposing abortion"*: No author listed, "Resolu-
tion passed by the Assembly—Affirming the Sanctity of Human Life, 60th
Annual Meeting, Association of American Physicians and Surgeons, Point
Clear, Alabama, September 17–19, 2003," Association of American Physicians and Surgeons, aapsonline.org/about-aaps/ (accessed November 25,
2019).

201 *"denying human involvement in climate change"*: Richard S. Lindzen, "Science in the Public Square: Global Climate Alarmism and Historical Prece-
dents," *Journal of American Physicians and Surgeons* 18, no. 3 (Fall 2013): 70.

201 *"denying that HIV causes AIDS"*: Henry H. Bauer, "Questioning HIV/
AIDS: Morally Reprehensible or Scientifically Warranted?," *Journal of
American Physicians and Surgeons* 12, no. 4 (Winter 2007): 116.

201 *"defending Rush Limbaugh"*: Debra Mullins, "The American Association
of Physicians and Surgeons Files Health Care Lawsuit," The Post & Email,
May 25, 2010, www.thepostemail.com/2010/05/25/the-american-association
-of-physicians-and-surgeons-files-health-care-lawsuit/ (accessed November
25, 2019).

201 *"claiming that Obama used 'neuro-linguistic programming'"*: Joseph
Gerth, "From the Archives: Paul in Group with Offbeat Views," *Courier-
Journal*, September 25, 2010, www.courier-journal.com/story/news/pol
itics/gerth/2015/02/04/rand-paul-in-association-of-american-physicians
-and-surgeons/22857153/ (accessed November 25, 2019).

202 *"the pro-family message of* King Lear*":* No author listed, "William

Shakespeare," www.conservapedia.com/William_Shakespeare (accessed November 25, 2019).

202 *"Phyllis Schlafly, who after all was four times more prolific"*: No author listed, "Phyllis Schlafly," www.conservapedia.com/Phyllis_Schlafly (accessed November 25, 2019).

202 *"'kids will be encouraged to practice sodomy in kindergarten'"*: Dylan Baddour, "Houston GOP Activist Steven Hotze: 'Kids Will Be Encouraged to Practice Sodomy in Kindergarten,'" *Houston Chronicle*, July 15, 2015, www .chron.com/news/politics/texas/article/Houston-GOP-activist-Steven-Hot ze-Kids-will-be-enc-6386471.php (accessed November 25, 2019).

202 *"the gay rights movement was backed by 'satanic cults'"*: Christopher Hooks, "Steve Hotze to Gay Nazis: En Garde!," *Texas Observer*, August 14, 2015, www.texasobserver.org/conservative-texas-steve-hotze-gay-nazis/ (accessed November 25, 2019).

202 *"'During my Senate confirmation hearings'"*: Craig Malisow, "Dr. Steven Hotze's Weird War Against the Texas Medical Board," *Houston Press*, March 26, 2014, www.houstonpress.com/news/dr-steven-hotzes-weird-war -against-the-texas-medical-board-6601127 (accessed November 25, 2019).

CHAPTER 19: *Edmund Husserl Would Like a Word*

209 *"for some reason food allergy could not"*: Smith, *Another Person's Poison*, 75.

209 *"whether food allergy should be regarded as 'real'"*: Smith, *Another Person's Poison*, 19.

209 *"'any form of altered biological reactivity'"*: Smith, *Another Person's Poison*, 19.

209 *"Theron Randolph . . . belonged to the latter group"*: Donna Chavez, "The Allergist That Roared," *The Chicago Tribune*, January 3, 1993, articles .chicagotribune.com/1993-01-03/features/9303150511_1_dr-randolph-allergy -clinic-theron-g-randolph (accessed November 25, 2019).

210 *"spent as many as four hours on a single intake"*: Douglas Hunt, *What Your Doctor May Not Tell You About Anxiety, Phobias, & Panic Attacks* (New York: Warner Books, 2005).

210 *"'I might be equally misinformed 10 years later'"*: Donna Chavez, "The Allergist That Roared," *The Chicago Tribune*, January 3, 1993, articles .chicagotribune.com/1993-01-03/features/9303150511_1_dr-randolph-allergy -clinic-theron-g-randolph (accessed November 25, 2019).

210 *"he was allergic to peanuts, maple, and corn"*: Theron G. Randolph,

Environmental Medicine: Beginnings & Bibliographies of Clinical Ecology, 56.

210 *"'a pernicious influence for medical students'"*: Donna Chavez, "The Allergist That Roared," *The Chicago Tribune,* January 3, 1993, articles.chicago tribune.com/1993-01-03/features/9303150511_1_dr-randolph-allergy -clinic-theron-g-randolph (accessed November 25, 2019).

210 *"displacing the experience of individual patients with the generalized results"*: J. Fuller, "The New Medical Model: A Renewed Challenge for Biomedicine," *Canadian Medical Association Journal* 189, no. 17 (May 1, 2017): E640.

212 *"'remain attentive to the* phenomenon'": Fred Dallmayr, "Heidegger and Freud," *Political Psychology* 14, no. 2 (June 1993): 241.

212 *"'We hear the column on the march, the north wind'"*: Martin Heidegger, *Being and Time,* Joan Stambaugh, trans. (Albany: State University of New York Press, 1996), 153.

213 *"'There is the highest need for doctors who think'"*: Kevin Aho, *Existential Medicine: Essays on Health and Illness* (Lanham: Rowman & Littlefield International, 2018), xii.

213 *"Heidegger agreed to give a series of seminars"*: Aho, *Existential Medicine,* xi.

213 *"Heidegger chalking a semicircle . . . on the blackboard"*: Rüdiger Safranski, *Martin Heidegger: Between Good and Evil* (Cambridge: Harvard University Press, 1999), 405.

213 *"A person might blush for a handful of reasons"*: Hermann Lang, Stefan Brunnhuber, and F. Rudolph, "The So-Called Zollikon Seminars: Heidegger as a Psychotherapist," *Journal of American Academy of Psychoanalysis and Dynamic Psychiatry* 32, no. 2 (Summer 2003): 352.

213 *"'I do not* have *a body'"*: Maria Lucia Araujo Sadala, Ph.D., R.G.N., "Phenomenology as a Method to Investigate the Experience Lived: A Perspective from Husserl and Merleau Ponty's Thought," *Journal of Advanced Nursing* 37, no. 3 (February 2002): 286.

214 *"The same violence committed in an immersive virtual reality"*: Joshua Rothman, "Are We Already Living in Virtual Reality?," *The New Yorker,* April 2, 2018 (accessed November 25, 2018).

214 *"'to remember that our patients are people'"*: Richard J. Baron, M.D., "An Introduction to Medical Phenomenology: I Can't Hear You While I'm Listening," *Annals of Internal Medicine* 103, no. 4 (October 1985).

214 *"'embrace a compassionate attitude that shares existentially'"*: James A. Marcum, "Reflections on Humanizing Biomedicine," *Perspectives in Biology and Medicine* 51, no. 3 (Summer 2008): 399.

215 *"toward a more contextualized understanding of 'affect'"*: Robert D. Stolorow, "The Renewal of Humanism in Psychoanalytic Therapy," *Psychotherapy* 49, no. 4 (December 2012): 442.

215 *"Heidegger's ideas received the warmest reception from nurses"*: O. P. Wiggins and M. A. Schwartz, "Richard Zaner's Phenomenology of the Clinical Encounter," *Theoretical Medicine and Bioethics* 26, no. 1 (2005); M. H. Wilde, "Embodied Knowledge in Chronic Illness and Injury," *Nursing Inquiry* 10, no. 3 (September 2003); Maura Dowling, "From Husserl to van Manen: A Review of Different Phenomenological Approaches," *International Journal of Nursing Studies* 44, no. 1 (January 2007).

215 *"It was also 95 percent female"*: Emily Rappleye, "Gender Ratio of Nurses Across 50 States," Becker's Hospital Review, May 29, 2015, www.beckers hospitalreview.com/human-capital-and-risk/gender-ratio-of-nurses-across -50-states.html (accessed November 25, 2019).

215 *"communication between doctors and nurses was terrible"*: Milisa Manojlovich et al., "Formative Evaluation of the Video Reflexive Ethnography Method, as Applied to the Physician–Nurse Dyad," *BMJ Quality and Safety* 28, no. 2 (2019): 160.

215 *"increasingly borne out by advances in neuroscience"*: Firat Soylu, "An Embodied Approach to Understanding: Making Sense of the World Through Simulated Bodily Activity," *Frontiers in Psychology* 7 (December 2016); S. M. Rosen, "Why Natural Science Needs Phenomenological Philosophy," *Progress in Biophysics and Molecular Biology* 119, no. 3 (December 2015); M. J. Schroeder and J. Vallverdú, "Situated Phenomenology and Biological Systems: Eastern and Western Synthesis," *Progress in Biophysics and Molecular Biology* 119, no. 3 (December 2015); P. C. Marijuán, J. Navarro, and R. del Moral, "How the Living Is in the World: An Inquiry into the Informational Choreographies of Life," *Progress in Biophysics and Molecular Biology* 119, no. 3 (December 2015).

215 *"(. . . semantic understanding is at least partly grounded in physical experience)"*: C. Galetzka, "The Story So Far: How Embodied Cognition Advances Our Understanding of Meaning-Making," *Frontiers in Psychology* 8 (July 2017): 2.

215 *"while preparing for a class on Descartes"*: Ted Toadvine, "Maurice Merleau-Ponty," Stanford Encyclopedia of Philosophy, September 14, 2016, plato.stanford.edu/entries/merleau-ponty/ (accessed November 25, 2019).

CHAPTER 20: *What You Can Learn from a Butterfly*

218   *"learned to transmute it into an asthma attack"*: Bessel van der Kolk, *The Body Keeps the Score: Brain, Mind, and Body in the Healing of Trauma* (New York: Penguin Books, 2014), 99.

218   *"migraines, chronic back and neck pain, fibromyalgia, digestive problems"*: van der Kolk, *The Body Keeps the Score,* 100.

218   *"'They learned to shut down their once overwhelming emotions'"*: van der Kolk, *The Body Keeps the Score,* 101.

218   *"In children in particular, simple neglect was enough"*: van der Kolk, *The Body Keeps the Score,* 59, 90.

219   *"'Children are . . . programmed to be fundamentally loyal'"*: van der Kolk, *The Body Keeps the Score,* 135.

220   *"began his career in the 1940s as a hard-core biomedicine guy"*: Jules Cohen and Stephanie Brown Clark, *John Romano & George Engel: Their Lives & Work* (Rochester, NY: University of Rochester Press, 2010), 14.

220   *"One summer he performed over three hundred autopsies"*: R. Ader and A. H. Schmale Jr., "George Libman Engel: On the Occasion of His Retirement," *Psychosomatic Medicine* 42, 1 Suppl. (1980): 81.

220   *"the eponysterical Soma Weiss"*: Cohen and Clark, *John Romano & George Engel,* 73.

220–221   *"he observed the way he pulled up a chair"*: Cohen and Clark, *John Romano & George Engel,* 75.

221   *"whether any get well cards were taped to the wall"*: Cohen and Clark, *John Romano & George Engel,* 124.

221   *"how the patient's answers made the student feel"*: George L. Engel and William L. Morgan, *Interviewing the Patient* (Philadelphia: W. B. Saunders, 1973), 6.

221   *"'interested . . . in human beings generally'"*: Cohen and Clark, *John Romano & George Engel,* 70.

221   *"a rousing call for 'a new medical model'"*: George L. Engel, "The Need for a New Medical Model: A Challenge for Biomedicine," *Science* 196, no. 4286 (April 8, 1977): 1331.

222   *"described his diagnostic skills as 'almost mythical'"*: Jeffrey P. Spike, "The Philosophy of George Engel and the Philosophy of Medicine," *Philosophy, Psychiatry & Psychology* 14, no. 4, (December 2007): 319.

222   *"'attentive to the patient's spontaneous associations'"*: Engel and Morgan, *Interviewing the Patient,* 39.

222 *"Michael Polanyi . . . had argued much the same thing"*: Michael Polyani, *The Tacit Dimension* (Garden City, NY: Doubleday & Co., 1966), 4.

222 *"(who was deeply influenced by Merleau-Ponty)"*: M. H. Wilde, "Embodied Knowledge in Chronic Illness and Injury," *Nursing Inquiry* 10, no. 3 (September 2003): 173.

222 *"Master to apprentice"*: Michael Polanyi, *Personal Knowledge: Towards a Post-Critical Philosophy* (Chicago: University of Chicago Press; Corr. edition, 1974), 53.

222 *"expressed regret that his landmark 1977 paper"*: Cohen and Clark, *John Romano & George Engel*, 181.

222 *"if it could not be called 'science' then it could only be called 'art,'"*: Cohen and Clark, *John Romano & George Engel*, 181.

222 *"a philosophically inclined Austrian biologist"*: Mark Davidson, *Uncommon Sense: The Life and Thought of Ludwig von Bertalanffy, Father of General Systems Theory* (New York: J. P. Tarcher, Inc., 1983), 49–52.

223 *"'one of those strategically placed thinkers'"*: Debora Hammond, *The Science of Synthesis: Exploring the Social Implications of General Systems Theory* (Boulder: University Press of Colorado, 2010), 104.

223 *"a gestalt response of the entire organism"*: Manfred Drack, Wilfried Apfalter, and David Pouvreau, "On the Making of a System Theory of Life: Paul A. Weiss and Ludwig von Bertalanffy's Conceptual Connection," *The Quarterly Review of Biology* 82, no. 4 (2007): 19.

223 *"'The elementary steps in behavior are subordinated to the state of the whole'"*: Peter H. Raven, "Paul Alfred Weiss." In *Biographical Memoirs, Volume 72* (1997), 373–86 (Washington, D.C.: National Academies Press, 1997).

224 *"'It is an urgent task for the future'"*: Manfred Drack and Wilfried Apfalter, "Is Paul A. Weiss' and Ludwig von Bertalanffy's System Thinking Still Valid Today?," *Systems Research and Behavioral Science* 24, no. 5 (November 2007): 540.

224 *"By the early 1930s he was calling for a new approach"*: Manfred Drack and Wilfried Apfalter, "Is Paul A. Weiss' and Ludwig von Bertalanffy's System Thinking Still Valid Today?," *Systems Research and Behavioral Science* 24, no. 5 (November 2007): 541.

224 *"'Classical science in its diverse disciplines'"*: Ludwig von Bertalanffy, "The Quest for Systems Philosophy," *Metaphilosophy* 3, no. 2 (April 1972): 142–45.

225 *"At the bottom you had complex proteins"*: Thaddus E. Weckowicz, "Ludwig von Bertalanffy (1901–1972): A Pioneer of General Systems Theory," Center for Systems Research Working Paper No. 89-2, 2000.

225 *"the whole was greater than the sum of its parts"*: Michael Polanyi, "Transcendence and Self-Transcendence," *Soundings: An Interdisciplinary Journal* 53, no. 1 (1970): 88–94.

226 *"Engel dubbed this 'the Hierarchy of Natural Systems'"*: George Engel, "The Biopsychosocial Model and Medical Education: Who Are to Be the Teachers?," *New England Journal of Medicine* 306, no. 13 (April 1, 1982): 802.

226 *"no alternative but to behave in a humane and empathic manner"*: George Engel, "How Much Longer Must Medicine's Science Be Bound by a Seventeenth Century World View?," *Psychotherapy and Psychosomatics* 57, nos. 1–2 (1992): 8.

226 *"cited over five hundred times"*: Database search: Web of Science. Parameters: Title: (Need for a New Medical Model—Challenge for Biomedicine); Timespan: 1977–1979; Databases: WOS, BCI, CCC, DRCI, DIIDW, KJD, MEDLINE, RSCI, SCIELO, ZOOREC, webofknowledge.com (accessed November 25, 2019).

226 *"the gender ratio was nearly as lopsided as in nursing"*: "Labor Force Statistics from the Current Population Survey: 11. Employed Persons by Detailed Occupation, Sex, Race, and Hispanic or Latino Ethnicity," United States Department of Labor, www.bls.gov/cps/cpsaat11.htm (accessed November 25, 2019).

226 *"wasn't science so much as a way to think about doing science"*: Ervin László, foreword to Ludwig von Bertalanffy, *Perspectives on General System Theory* (New York: George Braziller, 1976).

226 *"'permission to do everything, but no specific guidance'"*: S. Nassir Ghaemi, "The Biopsychosocial Model in Psychiatry: A Critique," *Existenz* 6, no. 1 (2011): 4.

226 *"'a list of ingredients, as opposed to a recipe'"*: Paul R. McHugh and Phillip R. Slavney, *The Perspectives of Psychiatry* (Baltimore: Johns Hopkins University Press, 1983; 2nd edition, 1998).

227 *"'what it feels like to be a serotonin molecule'"*: Bradley Lewis, "George Engel's Legacy for the Philosophy of Medicine and Psychiatry," *Philosophy, Psychiatry & Psychology* 14, no. 4 (December 2007): 329.

228 *"'the student identifies with either the psychiatrist or the internist'"*: Bruce S. Singh, "George Engel: A Personal Reminiscence," *Australian and New Zealand Journal of Psychiatry* 36, no. 4 (August 2002): 470.

228 *"poor predictors of 'subjective' health"*: Havi Carel, "Phenomenology and Its Application in Medicine," *Theoretical Medicine and Bioethics* 32, no. 1 (February 2011): 43.

228   *"organ-based medical specialties more entrenched than ever"*: Giovanni A. Favaa and Nicoletta Sonino, "From the Lesson of George Engel to Current Knowledge: The Biopsychosocial Model 40 Years Later," *Psychotherapy and Psychosomatics* 86, no. 5 (2017): 257.

228   *"82 percent of patients whose symptoms defy the dominant"*: W. Hiller, W. Rief, and E. Brähler, "Somatization in the Population: From Mild Bodily Misperceptions to Disabling Symptoms," *Social Psychiatry and Psychiatric Epidemiology* 41, no. 9 (September 2006): 710.

228   *"patients given eleven seconds on average to explain"*: Naykky Singh Ospina et al., "Eliciting the Patient's Agenda: Secondary Analysis of Recorded Clinical Encounters," *Journal of General Internal Medicine* 34, no. 1 (January 2019): 39.

228   *"burnout rates . . . recently increased 9 percentage points"*: Iqbal Pittalwala, "Three Factors Could Explain Physician Burnout," UCR Today, August 17, 2018, ucrtoday.ucr.edu/53530 (accessed November 25, 2019).

228   *"suicide rates were more than double that of the general population"*: Blake Farmer, "When Doctors Struggle with Suicide, Their Profession Often Fails Them," NPR, July 31, 2018, www.npr.org/sections/health-shots/2018/07/31/634217947/to-prevent-doctor-suicides-medical-industry-rethinks-how-doctors-work (accessed November 25, 2019).

CHAPTER 21: *Alone with Orcas*

231   *"million-dollar Sedona house"*: Estimate obtained from Zillow.com (Accessed November 25, 2019).

236   *"'can turn the whole world into a gathering of aliens'"*: van der Kolk, *The Body Keeps the Score*, 81.

CHAPTER 22: *The King of the Forest*

242   *"a 2012 degree from Tufts Medical School"*: Confidential source, Facebook.

243   *"'only at the bright edges of the world'"*: Willa Cather, *Death Comes for the Archbishop* (New York: Vintage Classics, 199), 273.

246   *"the idea of giving her a horse"*: van der Kolk, *The Body Keeps the Score*, 152.

CHAPTER 24: *Bernhoft's Judgment*

256 *"Just another freak in the freak kingdom"*: Hunter S. Thompson, *Fear and Loathing in Las Vegas: A Savage Journey to the Heart of the American Dream* (New York: Vintage, 1998), 83.

Index